S0-BKU-011

Mathematics

SIMPLIFIED AND SELF-TAUGHT

Mathematics

SIMPLIFIED AND SELF-TAUGHT

BARBARA ERDSNEKER, M.A.

ARCO PUBLISHING, INC.
NEW YORK

Sixth Edition, Second Printing, 1983

Published by Arco Publishing, Inc.
215 Park Avenue South, New York, N.Y. 10003

Library of Congress Cataloging in Publication Data
Erdsneker, Barbara.
Mathematics simplified and self-taught.
1. Mathematics—1961- 2. Mathematics—Examinations, questions, etc. I. Title.
QA39.2.E73 1982 513'.143 81-14912
ISBN 0-668-05357-7 AACR2

Printed in the United States of America

CONTENTS

What This Book Will Do for You vi

PART ONE—MATH MADE SIMPLE

Fractions 3
  Practice Problems Involving Fractions 10
  Problem Solutions—Fractions 11
Decimals 14
  Practice Problems Involving Decimals 19
  Problem Solutions—Decimals 20
Percents 23
  Practice Problems Involving Percents 26
  Problem Solutions—Percents 27
Shortcuts in Multiplication and Division 30
  Practice Problems 37
  Solutions to Practice Problems 38
Powers and Roots 39
  Practice Problems Involving Powers and Roots 42
  Problem Solutions—Powers and Roots 42
Table of Measures 44
Denominate Numbers (Measurement) 47
  Practice Problems Involving Measurement 52
  Problem Solutions—Measurement 53
Statistics and Probability 56
  Practice Problems Involving Statistics and Probability 58
  Problem Solutions—Statistics and Probability 59
Graphs 61
  Practice Problems Involving Graphs 65
  Problem Solutions—Graphs 69
Payroll 71
  Practice Problems Involving Payroll 73
  Payroll Problems—Solutions 75
Sequences 76
  Practice Problems Involving Sequences 78
  Problem Solutions—Sequences 79
Operations with Algebraic Expressions 80
  Practice Problems Involving Algebraic Expressions 91
  Problem Solutions—Algebraic Expressions 92
Equations, Inequalities and Problems in Algebra 94
  Practice Problems Involving Equations, Inequalities and Problems in Algebra 107
  Problem Solutions—Problems in Algebra 109
Geometry 114
  Practice Problems Involving Geometry 129
  Problem Solutions—Geometry 130
Coordinate Geometry 133
  Practice Problems Involving Graphs 138
  Problem Solutions—Graphs 139

PART TWO—THREE EXAMINATIONS FOR REVIEW

Examination I—Civil Service Arithmetic 145
Examination II—Mathematics for GED Candidates 157
Examination III—Mathematics for the SAT 169

WHAT THIS BOOK WILL DO FOR YOU

This book is based on a simple and perhaps unusual idea: that mathematics is a valuable tool in everyday living and that it is not difficult to learn.

Because MATHEMATICS SIMPLIFIED AND SELF-TAUGHT deals with the knowledge that is needed and used most often in business and personal affairs, it will be particularly easy to study and master. You will find that what you need to learn and want to learn, you will learn easily.

If you want to advance in your job or business, or in the Armed Forces, or get a better mark on Civil Service, High School Equivalency or College Entrance Examinations, you will find this book extremely helpful. It was written with your needs in mind, after a careful analysis of the kinds of mathematics needed in these areas.

It is suggested that you study the book systematically to acquaint yourself with all it contains and build up an organized body of mathematical knowledge. Part One is divided into twenty topics. Each topic is treated in a straightforward, step-by-step fashion presenting everything you need to know with numerous examples to illustrate the points made in the text. At the end of each section you will find practice problems to test your comprehension of the material you have just learned. Work out these problems on your own first. Then check your answers with the detailed solutions provided to see how well you are progressing. When you are satisfied that you understand a particular topic and you are able to answer seventy-five percent of the practice problems correctly, move on to the next topic and repeat the study — test yourself — rate your progress technique.

Part Two of MATHEMATICS SIMPLIFIED AND SELF-TAUGHT consists of three sample exams. The first exam is made up of problems similar to the ones that appear on Civil Service, military and employment tests. These problems represent the kind of practical mathematical knowledge required in business and everyday situations.

The second exam is patterned on the mathematics section of the High School Equivalency Diploma Test. Here you will find problems similar to the actual test in type of questions asked, areas covered and level of difficulty. It indicates the mathematical ability required of the average high school graduate.

The final examination in this book provides a sample of the mathematical knowledge expected of college-bound students who must take the Scholastic Aptitude Test (SAT). This exam is divided into two parts because the SAT actually contains at least two mathematical ability sections. It covers the entire range of mathematical knowledge necessary to achieve success on this examination.

Each sample examination is followed by solutions for every problem. It's a good idea to look over these solutions — even for the problems you answered correctly. The solutions provide the reasoning behind each answer which will help you to solve other similar problems.

Helpful Hint for Solving Multiple-Choice Questions in Mathematics

Always work out your own solution before checking the answer choices. The answer choices are usually too close to allow for mere estimates and most questions will include among the answer choices those answers most likely to be arrived at through common calculating errors or faulty mathematical reasoning.

Mathematics

SIMPLIFIED AND SELF-TAUGHT

Part One

MATH MADE SIMPLE

FRACTIONS

Fractions and Mixed Numbers

1. A **fraction** is part of a unit.

   a. A fraction has a **numerator** and a **denominator**.

   Example: In the fraction $\frac{3}{4}$, 3 is the numerator and 4 is the denominator.

   b. In any fraction, the numerator is being divided by the denominator.

   Example: The fraction $\frac{2}{7}$ indicates that 2 is being divided by 7.

   c. In a fraction problem, the whole quantity is 1, which may be expressed by a fraction in which the numerator and denominator are the same number.

   Example: If the problem involves $\frac{1}{8}$ of a quantity, then the whole quantity is $\frac{8}{8}$, or 1.

2. A **mixed number** is an integer together with a fraction, such as $2\frac{3}{5}$, $7\frac{3}{8}$, etc. The integer is the integral part, and the fraction is the fractional part.

3. An **improper fraction** is one in which the numerator is equal to or greater than the denominator, such as $\frac{19}{6}$, $\frac{25}{4}$, or $\frac{10}{10}$.

4. To change a mixed number to an improper fraction:

   a. Multiply the denominator of the fraction by the integer.

   b. Add the numerator to this product.

   c. Place this sum over the denominator of the fraction.

   Illustration: Change $3\frac{4}{7}$ to an improper fraction.

   SOLUTION:
   $7 \times 3 = 21$
   $21 + 4 = 25$
   $3\frac{4}{7} = \frac{25}{7}$

   Answer: $\frac{25}{7}$

5. To change an improper fraction to a mixed number:

   a. Divide the numerator by the denominator. The quotient, disregarding the remainder, is the integral part of the mixed number.

   b. Place the remainder, if any, over the denominator. This is the fractional part of the mixed number.

   Illustration: Change $\frac{36}{13}$ to a mixed number.

   SOLUTION:
   $$\begin{array}{r} 2 \\ 13\overline{)36} \\ \underline{26} \\ 10 \end{array} \text{ remainder}$$
   $\frac{36}{13} = 2\frac{10}{13}$

   Answer: $2\frac{10}{13}$

6. The numerator and denominator of a fraction may be changed by multiplying both by the same number, without affecting the value of the fraction.

   Example: The value of the fraction $\frac{2}{5}$ will not be altered if the numerator and the denominator are multiplied by 2, to result in $\frac{4}{10}$.

7. The numerator and the denominator of a fraction may be changed by dividing both by the same number, without affecting the value of the fraction. This process is called **reducing the fraction**. A fraction that has been reduced as much as possible is said to be in **lowest terms**.

   Example: The value of the fraction $\frac{3}{12}$ will not be altered if the numerator and denominator are divided by 3, to result in $\frac{1}{4}$.

   Example: If $\frac{6}{30}$ is reduced to lowest terms (by dividing both numerator and denominator by 6), the result is $\frac{1}{5}$.

8. As a final answer to a problem:

   a. Improper fractions should be changed to mixed numbers.

   b. Fractions should be reduced as far as possible.

Addition of Fractions

9. Fractions cannot be added unless the denominators are all the same.

   a. If the denominators are the same, add all the numerators and place this sum over the common denominator. In the case of mixed numbers, follow the above rule for the fractions and then add the integers.

   Example: The sum of $2\frac{3}{8} + 3\frac{1}{8} + \frac{3}{8} = 5\frac{7}{8}$.

   b. If the denominators are not the same, the fractions, in order to be added, must be converted to ones having the same denominator. To do this, it is first necessary to find the lowest common denominator.

10. The **lowest common denominator** (henceforth called the L.C.D.) is the lowest number that can be divided evenly by all the given denominators. If no two of the given denominators can be divided by the same number, then the L.C.D. is the product of all the denominators.

    Example: The L.C.D. of $\frac{1}{2}$, $\frac{1}{3}$, and $\frac{1}{5}$ is $2 \times 3 \times 5 = 30$.

11. To find the L.C.D. when two or more of the given denominators can be divided by the same number:

    a. Write down the denominators, leaving plenty of space between the numbers.

    b. Select the smallest number (other than 1) by which one or more of the denominators can be divided evenly.

    c. Divide the denominators by this number, copying down those that cannot be divided evenly. Place this number to one side.

    d. Repeat this process, placing each divisor to one side until there are no longer any denominators that can be divided evenly by any selected number.

e. Multiply all the divisors to find the L.C.D.

Illustration: Find the L.C.D. of $\frac{1}{5}$, $\frac{1}{7}$, $\frac{1}{10}$, and $\frac{1}{14}$.

SOLUTION:

$$\begin{array}{r|rrrr} 2 & 5 & 7 & 10 & 14 \\ \hline 5 & 5 & 7 & 5 & 7 \\ \hline 7 & 1 & 7 & 1 & 7 \\ \hline & 1 & 1 & 1 & 1 \end{array}$$

$7 \times 5 \times 2 = 70$

Answer: The L.C.D. is 70.

12. To add fractions having different denominators:

    a. Find the L.C.D. of the denominators.

    b. Change each fraction to an equivalent fraction having the L.C.D. as its denominator.

    c. When all of the fractions have the same denominator, they may be added, as in the example following item 9a.

Illustration: Add $\frac{1}{4}$, $\frac{3}{10}$, and $\frac{2}{5}$.

SOLUTION: Find the L.C.D.:

$$\begin{array}{r|rrr} 2 & 4 & 10 & 5 \\ \hline 2 & 2 & 5 & 5 \\ \hline 5 & 1 & 5 & 5 \\ \hline & 1 & 1 & 1 \end{array}$$

L.C.D. $= 2 \times 2 \times 5 = 20$

$$\begin{array}{rcr} \frac{1}{4} & = & \frac{5}{20} \\ \frac{3}{10} & = & \frac{6}{20} \\ +\ \frac{2}{5} & = & +\ \frac{8}{20} \\ \hline & & \frac{19}{20} \end{array}$$

Answer: $\frac{19}{20}$

13. To add mixed numbers in which the fractions have different denominators, add the fractions by following the rules in item 12 above, then add the integers.

Illustration: Add $2\frac{5}{7}$, $5\frac{1}{2}$, and 8.

SOLUTION: L.C.D. = 14

$$\begin{array}{rcr} 2\frac{5}{7} & = & 2\frac{10}{14} \\ 5\frac{1}{2} & = & 5\frac{7}{14} \\ +\ 8 & = & +\ 8 \\ \hline & & 15\frac{17}{14} = 16\frac{3}{14} \end{array}$$

Answer: $16\frac{3}{14}$

Subtraction of Fractions

14. a. Unlike addition, which may involve adding more than two numbers at the same time, subtraction involves only two numbers.

    b. In subtraction, as in addition, the denominators must be the same.

15. To subtract fractions:

    a. Find the L.C.D.

    b. Change both fractions so that each has the L.C.D. as the denominator.

    c. Subtract the numerator of the second fraction from the numerator of the first, and place this difference over the L.C.D.

    d. Reduce, if possible.

    Illustration: Find the difference of $\frac{5}{8}$ and $\frac{1}{4}$.

    SOLUTION: L.C.D. = 8

$$\begin{array}{rcr} \frac{5}{8} & = & \frac{5}{8} \\ -\frac{1}{4} & = & -\frac{2}{8} \\ \hline & & \frac{3}{8} \end{array}$$

    Answer: $\frac{3}{8}$

16. To subtract mixed numbers:

    a. It may be necessary to "borrow," so that the fractional part of the first term is larger than the fractional part of the second term.

    b. Subtract the fractional parts of the mixed numbers and reduce.

    c. Subtract the integers.

    Illustration: Subtract $16\frac{4}{5}$ from $29\frac{1}{3}$.

    SOLUTION: L.C.D. = 15

$$\begin{array}{rcr} 29\frac{1}{3} & = & 29\frac{5}{15} \\ -\ 16\frac{4}{5} & = & -\ 16\frac{12}{15} \\ \hline \end{array}$$

    Note that $\frac{5}{15}$ is less than $\frac{12}{15}$. Borrow 1 from 29, and change to $\frac{15}{15}$.

$$\begin{array}{rcr} 29\frac{5}{15} & = & 28\frac{20}{15} \\ -\ 16\frac{12}{15} & = & -\ 16\frac{12}{15} \\ \hline & & 12\frac{8}{15} \end{array}$$

    Answer: $12\frac{8}{15}$

Multiplication of Fractions

17. a. To be multiplied, fractions need not have the same denominators.

    b. A whole number has the denominator 1 understood.

18. To multiply fractions:

    a. Change the mixed numbers, if any, to improper fractions.

    b. Multiply all the numerators, and place this product over the product of the denominators.

    c. Reduce, if possible.

Illustration: Multiply $\frac{2}{3} \times 2\frac{4}{7} \times \frac{5}{9}$.

SOLUTION:

$$2\frac{4}{7} = \frac{18}{7}$$
$$\frac{2}{3} \times \frac{18}{7} \times \frac{5}{9} = \frac{180}{189}$$
$$= \frac{20}{21}$$

Answer: $\frac{20}{21}$

19. a. **Cancellation** is a device to facilitate multiplication. To cancel means to divide a numerator and a denominator by the same number in a multiplication problem.

Example: In the problem $\frac{4}{7} \times \frac{5}{6}$, the numerator 4 and the denominator 6 may be divided by 2.

$$\frac{\overset{2}{\cancel{4}}}{7} \times \frac{5}{\underset{3}{\cancel{6}}} = \frac{10}{21}$$

b. The word "of" is often used to mean "multiply."

Example: $\frac{1}{2}$ of $\frac{1}{2} = \frac{1}{2} \times \frac{1}{2} = \frac{1}{4}$

20. To multiply a whole number by a mixed number:

    a. Multiply the whole number by the fractional part of the mixed number.

    b. Multiply the whole number by the integral part of the mixed number.

    c. Add both products.

Illustration: Multiply $23\frac{3}{4}$ by 95.

SOLUTION:

$$\frac{95}{1} \times \frac{3}{4} = \frac{285}{4}$$
$$= 71\frac{1}{4}$$
$$95 \times 23 = 2185$$
$$2185 + 71\frac{1}{4} = 2256\frac{1}{4}$$

Answer: $2256\frac{1}{4}$

Division of Fractions

21. The **reciprocal** of a fraction is that fraction inverted.

    a. When a fraction is inverted, the numerator becomes the denominator and the denominator becomes the numerator.

Example: The reciprocal of $\frac{3}{8}$ is $\frac{8}{3}$.

Example: The reciprocal of $\frac{1}{3}$ is $\frac{3}{1}$, or simply 3.

b. Since every whole number has the denominator 1 understood, the reciprocal of a whole number is a fraction having 1 as the numerator and the number itself as the denominator.

Example: The reciprocal of 5 (expressed fractionally as $\frac{5}{1}$) is $\frac{1}{5}$.

22. To divide fractions:

  a. Change all the mixed numbers, if any, to improper fractions.

  b. Invert the second fraction and multiply.

  c. Reduce, if possible.

Illustration: Divide $\frac{2}{3}$ by $2\frac{1}{4}$.

SOLUTION:
$$2\tfrac{1}{4} = \tfrac{9}{4}$$
$$\tfrac{2}{3} \div \tfrac{9}{4} = \tfrac{2}{3} \times \tfrac{4}{9}$$
$$= \tfrac{8}{27}$$

Answer: $\frac{8}{27}$

23. A **complex fraction** is one that has a fraction as the numerator, or as the denominator, or as both.

Example: $\dfrac{\frac{2}{3}}{5}$ is a complex fraction.

24. To clear (simplify) a complex fraction:

  a. Divide the numerator by the denominator.

  b. Reduce, if possible.

Illustration: Clear $\dfrac{\frac{3}{7}}{\frac{5}{14}}$.

SOLUTION:
$$\tfrac{3}{7} \div \tfrac{5}{14} = \tfrac{3}{7} \times \tfrac{14}{5} = \tfrac{42}{35}$$
$$= \tfrac{6}{5}$$
$$= 1\tfrac{1}{5}$$

Answer: $1\frac{1}{5}$

Comparing Fractions

25. If two fractions have the same denominator, the one having the larger numerator is the greater fraction.

Example: $\frac{3}{7}$ is greater than $\frac{2}{7}$.

26. If two fractions have the same numerator, the one having the larger denominator is the smaller fraction.

Example: $\frac{5}{12}$ is smaller than $\frac{5}{11}$.

27. To compare two fractions having different numerators and different denominators:

  a. Change the fractions to equivalent fractions having their L.C.D. as their new denominator.

b. Compare, as in the example following item 25.

Illustration: Compare $\frac{4}{7}$ and $\frac{5}{8}$.

SOLUTION: L.C.D. = 7 × 8 = 56

$\frac{4}{7} = \frac{32}{56}$

$\frac{5}{8} = \frac{35}{56}$

Answer: Since $\frac{35}{56}$ is larger than $\frac{32}{56}$, $\frac{5}{8}$ is larger than $\frac{4}{7}$.

Fraction Problems

28. Most fraction problems can be arranged in the form: "What fraction of a number is another number?" This form contains three important parts:

- The fractional part
- The number following "of"
- The number following "is"

a. If the fraction and the "of" number are given, multiply them to find the "is" number.

Illustration: What is $\frac{3}{4}$ of 20?

SOLUTION: Write the question as "$\frac{3}{4}$ of 20 is what number?" Then multiply the fraction $\frac{3}{4}$ by the "of" number, 20:

$$\frac{3}{\cancel{4}_1} \times \cancel{20}^{5} = 15$$

Answer: 15

b. If the fractional part and the "is" number are given, divide the "is" number by the fraction to find the "of" number.

Illustration: $\frac{4}{5}$ of what number is 40?

SOLUTION: To find the "of" number, divide 40 by $\frac{4}{5}$:

$$40 \div \frac{4}{5} = \frac{\cancel{40}^{10}}{1} \times \frac{5}{\cancel{4}_1}$$
$$= 50$$

Answer: 50

c. To find the fractional part when the other two numbers are known, divide the "is" number by the "of" number.

Illustration: What part of 12 is 9?

SOLUTION: $9 \div 12 = \frac{9}{12}$

$= \frac{3}{4}$

Answer: $\frac{3}{4}$

Practice Problems Involving Fractions

1. Reduce to lowest terms: $\frac{60}{108}$.
   (A) $\frac{1}{48}$ (C) $\frac{5}{9}$
   (B) $\frac{1}{3}$ (D) $\frac{10}{18}$

2. Change $\frac{27}{7}$ to a mixed number.
   (A) $2\frac{1}{7}$ (C) $6\frac{1}{3}$
   (B) $3\frac{6}{7}$ (D) $7\frac{1}{2}$

3. Change $4\frac{2}{3}$ to an improper fraction.
   (A) $\frac{10}{3}$ (C) $\frac{14}{3}$
   (B) $\frac{11}{3}$ (D) $\frac{42}{3}$

4. Find the L.C.D. of $\frac{1}{6}$, $\frac{1}{10}$, $\frac{1}{18}$, and $\frac{1}{21}$.
   (A) 160 (C) 630
   (B) 330 (D) 1260

5. Add $16\frac{3}{8}$, $4\frac{4}{5}$, $12\frac{3}{4}$, and $23\frac{5}{6}$.
   (A) $57\frac{91}{120}$ (C) 58
   (B) $57\frac{1}{4}$ (D) 59

6. Subtract $27\frac{5}{14}$ from $43\frac{1}{6}$.
   (A) 15 (C) $15\frac{8}{21}$
   (B) 16 (D) $15\frac{17}{21}$

7. Multiply $17\frac{5}{8}$ by 128.
   (A) 2200 (C) 2356
   (B) 2305 (D) 2256

8. Divide $1\frac{2}{3}$ by $1\frac{1}{9}$.
   (A) $\frac{2}{3}$ (C) $1\frac{23}{27}$
   (B) $1\frac{1}{2}$ (D) 6

9. What is the value of $12\frac{1}{6} - 2\frac{3}{8} - 7\frac{2}{3} + 19\frac{3}{4}$?
   (A) 21 (C) $21\frac{1}{8}$
   (B) $21\frac{7}{8}$ (D) 22

10. Simplify the complex fraction $\frac{\frac{4}{9}}{\frac{2}{5}}$.
    (A) $\frac{1}{2}$ (C) $\frac{2}{5}$
    (B) $\frac{9}{10}$ (D) $1\frac{1}{9}$

11. Which fraction is largest?
    (A) $\frac{9}{16}$ (C) $\frac{5}{8}$
    (B) $\frac{7}{10}$ (D) $\frac{4}{5}$

12. One brass rod measures $3\frac{5}{16}$ inches long and another brass rod measures $2\frac{3}{4}$ inches long. Together their length is
    (A) $6\frac{9}{16}$ in. (C) $6\frac{1}{16}$ in.
    (B) $5\frac{1}{8}$ in. (D) $5\frac{1}{16}$ in.

13. The number of half-pound packages of tea that can be weighed out of a box that holds $10\frac{1}{2}$ lb. of tea is
    (A) 5 (C) $20\frac{1}{2}$
    (B) $10\frac{1}{2}$ (D) 21

14. If each bag of tokens weighs $5\frac{3}{4}$ pounds, how many pounds do 3 bags weigh?
    (A) $7\frac{1}{4}$ (C) $16\frac{1}{2}$
    (B) $15\frac{3}{4}$ (D) $17\frac{1}{4}$

15. During one week, a man traveled $3\frac{1}{2}$, $1\frac{1}{4}$, $1\frac{1}{6}$, and $2\frac{3}{8}$ miles. The next week he traveled $\frac{1}{4}$, $\frac{3}{8}$, $\frac{9}{16}$, $3\frac{1}{16}$, $2\frac{5}{8}$, and $3\frac{3}{16}$ miles. How many more miles did he travel the second week than the first week?
    (A) $1\frac{37}{48}$ (C) $1\frac{3}{4}$
    (B) $1\frac{1}{2}$ (D) 1

16. A certain type of board is sold only in lengths of multiples of 2 feet. The shortest board sold is 6 feet and the longest is 24 feet. A builder needs a large quantity of this type of board in $5\frac{1}{2}$-foot lengths. For minimum waste the lengths to be ordered should be
    (A) 6 ft (C) 22 ft
    (B) 12 ft (D) 24 ft

17. A man spent $\frac{15}{16}$ of his entire fortune in buying a car for $7500. How much money did he possess?
    (A) $6000 (C) $7000
    (B) $6500 (D) $8000

18. The population of a town was 54,000 in the last census. It has increased $\frac{2}{3}$ since then. Its present population is
    (A) 18,000 (C) 72,000
    (B) 36,000 (D) 90,000

19. If one third of the liquid contents of a can evaporates on the first day and three fourths of the remainder evaporates on the second day, the fractional part of the original contents remaining at the close of the second day is
(A) $\frac{5}{12}$ (C) $\frac{1}{6}$
(B) $\frac{7}{12}$ (D) $\frac{1}{2}$

20. A car is run until the gas tank is $\frac{1}{8}$ full. The tank is then filled to capacity by putting in 14 gallons. The capacity of the gas tank of the car is
(A) 14 gal (C) 16 gal
(B) 15 gal (D) 17 gal

Fraction Problems — Correct Answers

1. **(C)**	6. **(D)**	11. **(D)**	16. **(C)**
2. **(B)**	7. **(D)**	12. **(C)**	17. **(D)**
3. **(C)**	8. **(B)**	13. **(D)**	18. **(D)**
4. **(C)**	9. **(B)**	14. **(D)**	19. **(C)**
5. **(A)**	10. **(D)**	15. **(A)**	20. **(C)**

Problem Solutions — Fractions

1. Divide the numerator and denominator by 12:

$$\frac{60 \div 12}{108 \div 12} = \frac{5}{9}$$

One alternate method (there are several) is to divide the numerator and denominator by 6 and then by 2:

$$\frac{60 \div 6}{108 \div 6} = \frac{10}{18}$$
$$\frac{10 \div 2}{18 \div 2} = \frac{5}{9}$$

Answer: **(C)** $\frac{5}{9}$

2. Divide the numerator (27) by the denominator (7):

$$\begin{array}{r} 3 \\ 7\overline{)27} \\ \underline{21} \\ 6 \end{array}$$ remainder

$$\tfrac{27}{7} = 3\tfrac{6}{7}$$

Answer: **(B)** $3\frac{6}{7}$

3.
$$4 \times 3 = 12$$
$$12 + 2 = 14$$
$$4\tfrac{2}{3} = \tfrac{14}{3}$$

Answer: **(C)** $\frac{14}{3}$

4.
2)	6	10	18	21	(2 is a divisor of 6, 10, and 18)
3)	3	5	9	21	(3 is a divisor of 3, 9, and 21)
3)	1	5	3	7	(3 is a divisor of 3)
5)	1	5	1	7	(5 is a divisor of 5)
7)	1	1	1	7	(7 is a divisor of 7)
	1	1	1	1	

L.C.D. $= 2 \times 3 \times 3 \times 5 \times 7 = 630$

Answer: **(C)** 630

5. L.C.D. = 120

$$\begin{array}{rcr} 16\frac{3}{8} & = & 16\frac{45}{120} \\ 4\frac{4}{5} & = & 4\frac{96}{120} \\ 12\frac{3}{4} & = & 12\frac{90}{120} \\ +\ 23\frac{5}{6} & = & +\ 23\frac{100}{120} \\ \hline & & 55\frac{331}{120} = 57\frac{91}{120} \end{array}$$

Answer: **(A)** $57\frac{91}{120}$

6. L.C.D. = 42

$$\begin{array}{rcrcr} 43\frac{1}{6} & = & 43\frac{7}{42} & = & 42\frac{49}{42} \\ -\ 27\frac{5}{14} & = & -\ 27\frac{15}{42} & = & -\ 27\frac{15}{42} \\ \hline & & & & 15\frac{34}{42} = 15\frac{17}{21} \end{array}$$

Answer: **(D)** $15\frac{17}{21}$

7.

$$17\frac{5}{8} = \frac{141}{8}$$

$$\frac{141}{\cancel{8}_{1}} \times \frac{\cancel{128}^{16}}{1} = 2256$$

Answer: **(D)** 2256

8.

$$1\frac{2}{3} \div 1\frac{1}{9} = \frac{5}{3} \div \frac{10}{9}$$

$$= \frac{\cancel{5}^{1}}{\cancel{3}_{1}} \times \frac{\cancel{9}^{3}}{\cancel{10}_{2}}$$

$$= \frac{3}{2}$$

$$= 1\frac{1}{2}$$

Answer: **(B)** $1\frac{1}{2}$

9. L.C.D. = 24

$$\begin{array}{rcrcr} 12\frac{1}{6} & = & 12\frac{4}{24} & = & 11\frac{28}{24} \\ -\ 2\frac{3}{8} & = & -\ 2\frac{9}{24} & = & -\ 2\frac{9}{24} \\ \hline & & & & 9\frac{19}{24} \end{array}$$

$$\begin{array}{rcr} 9\frac{19}{24} & = & 9\frac{19}{24} \\ -\ 7\frac{2}{3} & = & -\ 7\frac{16}{24} \\ \hline & & 2\frac{3}{24} \end{array}$$

$$\begin{array}{rcr} 2\frac{3}{24} & = & 2\frac{3}{24} \\ +\ 19\frac{3}{4} & = & +\ 19\frac{18}{24} \\ \hline & & 21\frac{21}{24} \end{array}$$

$$21\frac{21}{24} = 21\frac{7}{8}$$

Answer: **(B)** $21\frac{7}{8}$

10. To simplify a complex fraction, divide the numerator by the denominator:

$$\frac{4}{9} \div \frac{2}{5} = \frac{\cancel{4}^{2}}{9} \times \frac{5}{\cancel{2}_{1}}$$

$$= \frac{10}{9}$$

$$= 1\frac{1}{9}$$

Answer: **(D)** $1\frac{1}{9}$

11. Write all of the fractions with the same denominator. L.C.D. = 80

$$\frac{9}{16} = \frac{45}{80}$$

$$\frac{7}{10} = \frac{56}{80}$$

$$\frac{5}{8} = \frac{50}{80}$$

$$\frac{4}{5} = \frac{64}{80}$$

Answer: **(D)** $\frac{4}{5}$

12.

$$\begin{array}{rcr} 3\frac{5}{16} & = & 3\frac{5}{16} \\ +\ 2\frac{3}{4} & = & +\ 2\frac{12}{16} \\ \hline & & 5\frac{17}{16} \\ & = & 6\frac{1}{16} \end{array}$$

Answer: **(C)** $6\frac{1}{16}$ in.

13.

$$10\frac{1}{2} \div \frac{1}{2} = \frac{21}{2} \div \frac{1}{2}$$

$$= \frac{21}{\cancel{2}_{1}} \times \frac{\cancel{2}^{1}}{1}$$

$$= 21$$

Answer: **(D)** 21

14.

$$5\frac{3}{4} \times 3 = \frac{23}{4} \times \frac{3}{1}$$

$$= \frac{69}{4}$$

$$= 17\frac{1}{4}$$

Answer: **(D)** $17\frac{1}{4}$

15. First week:
L.C.D. = 24

$$\begin{array}{rcrl} 3\frac{1}{2} & = & 3\frac{12}{24} & \text{miles} \\ 1\frac{1}{4} & = & 1\frac{6}{24} & \\ 1\frac{1}{6} & = & 1\frac{4}{24} & \\ +\ 2\frac{3}{8} & = & +\ 2\frac{9}{24} & \\ \hline & & 7\frac{31}{24} & = 8\frac{7}{24}\text{ miles} \end{array}$$

Second week:
L.C.D. = 16

$$\begin{array}{rcrl} \frac{1}{4} & = & \frac{4}{16} & \text{miles} \\ \frac{3}{8} & = & \frac{6}{16} & \\ \frac{9}{16} & = & \frac{9}{16} & \\ 3\frac{1}{16} & = & 3\frac{1}{16} & \\ 2\frac{5}{8} & = & 2\frac{10}{16} & \\ +\ 3\frac{3}{16} & = & +\ 3\frac{3}{16} & \\ \hline & & 8\frac{33}{16} & = 10\frac{1}{16}\text{ miles} \end{array}$$

L.C.D. = 48

$$\begin{array}{rcrl} 10\frac{1}{16} & = & 9\frac{51}{48} & \text{miles second week} \\ -\ 8\frac{7}{24} & = & -\ 8\frac{14}{48} & \text{miles first week} \\ \hline & & 1\frac{37}{48} & \text{miles more traveled} \end{array}$$

Answer: **(A)** $1\frac{37}{48}$

16. Consider each choice:

Each 6-ft board yields one $5\frac{1}{2}$-ft board with $\frac{1}{2}$ ft waste.
Each 12-ft board yields two $5\frac{1}{2}$-ft boards with 1 ft waste. ($2 \times 5\frac{1}{2} = 11$; $12 - 11 = 1$ ft waste)
Each 24-ft board yields four $5\frac{1}{2}$-ft boards with 2 ft waste. ($4 \times 5\frac{1}{2} = 22$; $24 - 22 = 2$ ft waste)
Each 22 ft board may be divided into four $5\frac{1}{2}$-ft boards with no waste. ($4 \times 5\frac{1}{2} = 22$ exactly)

Answer: **(C)** 22 ft

17. $\frac{15}{16}$ of fortune is \$7500.
Therefore, his fortune $= 7500 \div \frac{15}{16}$

$$= \frac{\overset{500}{\cancel{7500}}}{1} \times \frac{16}{\underset{1}{\cancel{15}}}$$

$$= 8000$$

Answer: **(D)** \$8000

18. $\frac{2}{3}$ of 54,000 = increase

$$\text{Increase} = \frac{2}{\underset{1}{\cancel{3}}} \times \overset{18{,}000}{\cancel{54{,}000}}$$

$$= 36{,}000$$

$$\text{Present population} = 54{,}000 + 36{,}000$$

$$= 90{,}000$$

Answer: **(D)** 90,000

19. First day: $\frac{1}{3}$ evaporates
$\frac{2}{3}$ remains

Second day: $\frac{3}{4}$ of $\frac{2}{3}$ evaporates
$\frac{1}{4}$ of $\frac{2}{3}$ remains

The amount remaining is

$$\frac{1}{\underset{2}{\cancel{4}}} \times \frac{\overset{1}{\cancel{2}}}{3} = \frac{1}{6} \text{ of original contents}$$

Answer: **(C)** $\frac{1}{6}$

20. $\frac{7}{8}$ of capacity = 14 gal
therefore, capacity $= 14 \div \frac{7}{8}$

$$= \frac{\overset{2}{\cancel{14}}}{1} \times \frac{8}{\underset{1}{\cancel{7}}}$$

$$= 16 \text{ gal}$$

Answer: **(C)** 16 gal

DECIMALS

1. A **decimal**, which is a number with a decimal point (.), is actually a fraction, the denominator of which is understood to be 10 or some power of 10.

   a. The number of digits, or places, after a decimal point determines which power of 10 the denominator is. If there is one digit, the denominator is understood to be 10; if there are two digits, the denominator is understood to be 100, etc.

   Example: $.3 = \frac{3}{10}$, $.57 = \frac{57}{100}$, $.643 = \frac{643}{1000}$

   b. The addition of zeros after a decimal point does not change the value of the decimal. The zeros may be removed without changing the value of the decimal.

   Example: .7 = .70 = .700 and vice versa, .700 = .70 = .7

   c. Since a decimal point is understood to exist after any whole number, the addition of any number of zeros after such a decimal point does not change the value of the number.

   Example: 2 = 2.0 = 2.00 = 2.000

Addition of Decimals

2. Decimals are added in the same way that whole numbers are added, with the provision that the decimal points must be kept in a vertical line, one under the other. This determines the place of the decimal point in the answer.

   Illustration: Add 2.31, .037, 4, and 5.0017

   SOLUTION:

   $$\begin{array}{r} 2.3100 \\ .0370 \\ 4.0000 \\ +\ 5.0017 \\ \hline 11.3487 \end{array}$$

   Answer: 11.3487

Subtraction of Decimals

3. Decimals are subtracted in the same way that whole numbers are subtracted, with the provision that, as in addition, the decimal points must be kept in a vertical line, one under the other. This determines the place of the decimal point in the answer.

   Illustration: Subtract 4.0037 from 15.3

   SOLUTION:

   $$\begin{array}{r} 15.3000 \\ -\ \ 4.0037 \\ \hline 11.2963 \end{array}$$

   Answer: 11.2963

Multiplication of Decimals

4. Decimals are multiplied in the same way that whole numbers are multiplied.

   a. The number of decimal places in the product equals the sum of the decimal places in the multiplicand and in the multiplier.

   b. If there are fewer places in the product than this sum, then a sufficient number of zeros must be added in front of the product to equal the number of places required, and a decimal point is written in front of the zeros.

Illustration: Multiply 2.372 by .012

SOLUTION:

```
   2.372    (3 decimal places)
×   .012    (3 decimal places)
 -------
    4744
   2372
 -------
 .028464    (6 decimal places)
```

Answer: .028464

5. A decimal can be multiplied by a power of 10 by moving the decimal point to the *right* as many places as indicated by the power. If multiplied by 10, the decimal point is moved one place to the right; if multiplied by 100, the decimal point is moved two places to the right; etc.

Example:
.235 × 10 = 2.35
.235 × 100 = 23.5
.235 × 1000 = 235

Division of Decimals

6. There are four types of division involving decimals:
   - When the dividend only is a decimal.
   - When the divisor only is a decimal.
   - When both are decimals.
   - When neither dividend nor divisor is a decimal.

   a. When the dividend only is a decimal, the division is the same as that of whole numbers, except that a decimal point must be placed in the quotient exactly above that in the dividend.

Illustration: Divide 12.864 by 32

SOLUTION:

```
       .402
32 ) 12.864
     12 8
     ----
        64
        64
```

Answer: .402

b. When the divisor only is a decimal, the decimal point in the divisor is omitted and as many zeros are placed to the right of the dividend as there were decimal places in the divisor.

Illustration: Divide 211327 by 6.817

```
                                         31000
SOLUTION:  6.817 ) 211327 = 6817 ) 211327000
          (3 decimal places)        20451         (3 zeros added)
                                     6817
                                     6817
```

Answer: 31000

c. When both divisor and dividend are decimals, the decimal point in the divisor is omitted and the decimal point in the dividend must be moved to the right as many decimal places as there were in the divisor. If there are not enough places in the dividend, zeros must be added to make up the difference.

Illustration: Divide 2.62 by .131

```
                                  20
SOLUTION:  .131 ) 2.62 = 131 ) 2620
                               262
```

Answer: 20

d. In instances when neither the divisor nor the dividend is a decimal, a problem may still involve decimals. This occurs in two cases: when the dividend is a smaller number than the divisor; and when it is required to work out a division to a certain number of decimal places. In either case, write in a decimal point after the dividend, add as many zeros as necessary, and place a decimal point in the quotient above that in the dividend.

Illustration: Divide 7 by 50.

```
               .14
SOLUTION:  50 ) 7.00
                5 0
                2 00
                2 00
```

Answer: .14

Illustration: How much is 155 divided by 40, carried out to 3 decimal places?

```
               3.875
SOLUTION:  40 ) 155.000
               120
                35 0
                32 0
                 3 00
                 2 80
                   200
```

Answer: 3.875

7. A decimal can be divided by a power of 10 by moving the decimal to the *left* as many places as indicated by the power. If divided by 10, the decimal point is moved one place to the left; if divided by 100, the decimal point is moved two places to the left; etc. If there are not enough places, add zeros in front of the number to make up the difference and add a decimal point.

Example: .4 divided by 10 = .04
.4 divided by 100 = .004

Rounding Decimals

8. To round a number to a given decimal place:

   a. Locate the given place.

   b. If the digit to the right is less than 5, omit all digits following the given place.

   c. If the digit to the right is 5 or more, raise the given place by 1 and omit all digits following the given place.

Examples: 4.27 = 4.3 to the nearest tenth
.71345 = .713 to the nearest thousandth

9. In problems involving money, answers are usually rounded to the nearest cent.

Conversion of Fractions to Decimals

10. A fraction can be changed to a decimal by dividing the numerator by the denominator and working out the division to as many decimal places as required.

Illustration: Change $\frac{5}{11}$ to a decimal of 2 places.

SOLUTION: $\frac{5}{11} = 11\overline{)5.00}$

$$\begin{array}{r} .45\frac{5}{11} \\ 11\,\overline{)\,5.00} \\ \underline{4.4} \\ 60 \\ \underline{55} \\ 5 \end{array}$$

Answer: $.45\frac{5}{11}$

11. To clear fractions containing a decimal in either the numerator or the denominator, or in both, divide the numerator by the denominator.

Illustration: What is the value of $\frac{2.34}{.6}$?

SOLUTION: $\frac{2.34}{.6} = .6\overline{)2.34} = 6\overline{)23.4}$

```
   3.9
6 ) 23.4
    18
    5 4
    5 4
```

Answer: 3.9

Conversion of Decimals to Fractions

12. Since a decimal point indicates a number having a denominator that is a power of 10, a decimal can be expressed as a fraction, the numerator of which is the number itself and the denominator of which is the power indicated by the number of decimal places in the decimal.

Example: $.3 = \frac{3}{10}$, $.47 = \frac{47}{100}$

13. When the decimal is a mixed number, divide by the power of 10 indicated by its number of decimal places. The fraction does not count as a decimal place.

Illustration: Change $.25\frac{1}{3}$ to a fraction.

SOLUTION: $.25\frac{1}{3} = 25\frac{1}{3} \div 100$
$= \frac{76}{3} \times \frac{1}{100}$
$= \frac{76}{300} = \frac{19}{75}$

Answer: $\frac{19}{75}$

14. When to change decimals to fractions:

a. When dealing with whole numbers, do not change the decimal.

Example: In the problem $12 \times .14$, it is better to keep the decimal:

$$12 \times .14 = 1.68$$

b. When dealing with fractions, change the decimal to a fraction.

Example: In the problem $\frac{3}{5} \times .17$, it is best to change the decimal to a fraction:

$$\frac{3}{5} \times .17 = \frac{3}{5} \times \frac{17}{100} = \frac{51}{500}$$

15. Because decimal equivalents of fractions are often used, it is helpful to be familiar with the most common conversions.

$\frac{1}{2} = .5$	$\frac{1}{3} = .3333$
$\frac{1}{4} = .25$	$\frac{2}{3} = .6667$
$\frac{3}{4} = .75$	$\frac{1}{6} = .1667$
$\frac{1}{5} = .2$	$\frac{1}{7} = .1429$
$\frac{1}{8} = .125$	$\frac{1}{9} = .1111$
$\frac{1}{16} = .0625$	$\frac{1}{12} = .0833$

Note that the left column contains exact values. The values in the right column have been rounded to the nearest ten-thousandth.

Practice Problems Involving Decimals

1. Add 37.03, 11.5627, 3.4005, 3423, and 1.141.
   (A) 3476.1342 (C) 3524.4322
   (B) 3500 (D) 3424.1342

2. Subtract 4.64324 from 7.
   (A) 3.35676 (C) 2.45676
   (B) 2.35676 (D) 2.36676

3. Multiply 27.34 by 16.943.
   (A) 463.22162 (C) 462.52162
   (B) 453.52162 (D) 462.53162

4. How much is 19.6 divided by 3.2, carried out to 3 decimal places?
   (A) 6.125 (C) 6.123
   (B) 6.124 (D) 5.123

5. What is $\frac{5}{11}$ in decimal form (to the nearest hundredth)?
   (A) .44 (C) .40
   (B) .55 (D) .45

6. What is $.64\frac{2}{3}$ in fraction form?
   (A) $\frac{97}{120}$ (C) $\frac{97}{130}$
   (B) $\frac{97}{150}$ (D) $\frac{98}{130}$

7. What is the difference between $\frac{3}{5}$ and $\frac{9}{8}$ expressed decimally?
   (A) .525 (C) .520
   (B) .425 (D) .500

8. A boy saved up $4.56 the first month, $3.82 the second month, and $5.06 the third month. How much did he save altogether?
   (A) $12.56 (C) $13.44
   (B) $13.28 (D) $14.02

9. The diameter of a certain rod is required to be $1.51 \pm .015$ inches. The rod would not be acceptable if the diameter measured
   (A) 1.490 in (C) 1.510 in
   (B) 1.500 in (D) 1.525 in

10. After an employer figures out an employee's salary of $190.57, he deducts $3.05 for social security and $5.68 for pension. What is the amount of the check after these deductions?
    (A) $181.84 (C) $181.93
    (B) $181.92 (D) $181.99

11. If the outer diameter of a metal pipe is 2.84 inches and the inner diameter is 1.94 inches, the thickness of the metal is
    (A) .45 in (C) 1.94 in
    (B) .90 in (D) 2.39 in

12. A boy earns $20.56 on Monday, $32.90 on Tuesday, $20.78 on Wednesday. He spends half of all that he earned during the three days. How much has he left?
    (A) $29.19 (C) $34.27
    (B) $31.23 (D) $37.12

13. The total cost of $3\frac{1}{2}$ pounds of meat at $1.69 a pound and 20 lemons at $.60 a dozen will be
    (A) $6.00 (C) $6.52
    (B) $6.40 (D) $6.92

14. A reel of cable weighs 1279 lb. If the empty reel weighs 285 lb and the cable weighs 7.1 lb per foot, the number of feet of cable on the reel is
    (A) 220 (C) 140
    (B) 180 (D) 100

15. 345 fasteners at $4.15 per hundred will cost
    (A) $ .1432 (C) $ 14.32
    (B) $1.4320 (D) $143.20

Decimal Problems — Correct Answers

1. (A)
2. (B)
3. (A)
4. (A)
5. (D)
6. (B)
7. (A)
8. (C)
9. (A)
10. (A)
11. (A)
12. (D)
13. (D)
14. (C)
15. (C)

Problem Solutions — Decimals

1. Line up all the decimal points one under the other. Then add:

```
     37.03
     11.5627
      3.4005
   3423.0000
 +    1.141
 -----------
   3476.1342
```

Answer: **(A)** 3476.1342

2. Add a decimal point and five zeros to the 7. Then subtract:

```
   7.00000
 - 4.64324
 ---------
   2.35676
```

Answer: **(B)** 2.35676

3. Since there are two decimal places in the multiplicand and three decimal places in the multiplier, there will be 2 + 3 = 5 decimal places in the product.

```
       27.34
   × 16.943
   ---------
        8202
      1 0936
     24 606
    164 04
    273 4
   ---------
   463.22162
```

Answer: **(A)** 463.22162

4. Omit the decimal point in the divisor by moving it one place to the right. Move the decimal point in the dividend one place to the right and add three zeros in order to carry your answer out to three decimal places, as instructed in the problem.

```
           6.125
 3.2. ) 19.6.000
        19 2
           4 0
           3 2
            80
            64
            160
            160
```

Answer: **(A)** 6.125

5. To convert a fraction to a decimal, divide the numerator by the denominator:

```
      .454
 11 ) 5.000
      4 4
        60
        55
         50
         44
          6
```

Answer: **(D)** .45 to the nearest hundredth

6. To convert a decimal to a fraction, divide by the power of 10 indicated by the number of decimal places. (The fraction does not count as a decimal place.)

$$64\tfrac{2}{3} \div 100 = \tfrac{194}{3} \div \tfrac{100}{1}$$
$$= \tfrac{194}{3} \times \tfrac{1}{100}$$
$$= \tfrac{194}{300}$$
$$= \tfrac{97}{150}$$

Answer: **(B)** $\tfrac{97}{150}$

7. Convert each fraction to a decimal and subtract to find the difference:

$\tfrac{9}{8} = 1.125$ $\tfrac{3}{5} = .60$

$$\begin{array}{r} 1.125 \\ -\ \ .60 \\ \hline .525 \end{array}$$

Answer: **(A)** .525

8. Add the savings for each month:

$$\begin{array}{r} \$4.56 \\ 3.82 \\ +\ \ 5.06 \\ \hline \$13.44 \end{array}$$

Answer: **(C)** $13.44

9.

$$\begin{array}{r} 1.51 \\ +\ \ .015 \\ \hline 1.525 \end{array} \qquad \begin{array}{r} 1.510 \\ -\ \ .015 \\ \hline 1.495 \end{array}$$

The rod may have a diameter of from 1.495 inches to 1.525 inches inclusive.

Answer: **(A)** 1.490 in.

10. Add to find total deductions:

$$\begin{array}{r} \$3.05 \\ +\ \ 5.68 \\ \hline \$8.73 \end{array}$$

Subtract total deductions from salary to find amount of check:

$$\begin{array}{r} \$190.57 \\ -\ \ \ 8.73 \\ \hline \$181.84 \end{array}$$

Answer: **(A)** $181.84

11. The difference of the two diameters equals the total thickness of the metal on both ends of the inner diameter.

$$\begin{array}{r} 2.84 \\ -1.94 \\ \hline .90 \end{array}$$

$.90 \div 2 = .45 =$ thickness of metal

Answer: **(A)** .45 in.

12. Add daily earnings to find total earnings:

$$\begin{array}{r} \$20.56 \\ 32.90 \\ +\ \ 20.78 \\ \hline \$74.24 \end{array}$$

Divide total earnings by 2 to find out what he has left:

$$\begin{array}{r} \$37.12 \\ 2\ \overline{)\ \$74.24} \end{array}$$

Answer: **(D)** $37.12

13. Find cost of $3\tfrac{1}{2}$ pounds of meat:

$$\begin{array}{r} \$1.69 \\ \times\ \ \ 3.5 \\ \hline 845 \\ 5\ 07 \\ \hline \$5.915 \end{array}$$

$5.915 = $5.92 to the nearest cent

Find cost of 20 lemons:

$\$.60 \div 12 = \$.05$ (for 1 lemon)
$\$.05 \times 20 = \$1.00$ (for 20 lemons)

Add cost of meat and cost of lemons:

$$\begin{array}{r} \$5.92 \\ +\ \ 1.00 \\ \hline \$6.92 \end{array}$$

Answer: **(D)** $6.92

14. Subtract weight of empty reel from total weight to find weight of cable:

$$\begin{array}{r} 1279\text{ lb} \\ -\ \ 285\text{ lb} \\ \hline 994\text{ lb} \end{array}$$

Each foot of cable weighs 7.1 lb. Therefore, to find the number of feet of cable on the reel, divide 994 by 7.1:

```
          14 0.
  7.1. ) 994.0.
         71
         284
         284
          0 0
```

Answer: **(C)** 140

15. Each fastener costs:

$$\$4.15 \div 100 = \$.0415$$

345 fasteners cost:

```
      345
  × .0415
     1725
     345
   13 80
  14.3175
```

Answer: **(C)** $14.32

PERCENTS

1. The **percent symbol** (%) means "parts of a hundred." Some problems involve expressing a fraction or a decimal as a percent. In other problems, it is necessary to express a percent as a fraction or a decimal in order to perform the calculations.

2. To change a whole number or a decimal to a percent:

   a. Multiply the number by 100.

   b. Affix a % sign.

   Illustration: Change 3 to a percent.

   SOLUTION: $3 \times 100 = 300$
   $3 = 300\%$

   Answer: 300%

   Illustration: Change .67 to a percent.

   SOLUTION: $.67 \times 100 = 67$
   $.67 = 67\%$

   Answer: 67%

3. To change a fraction or a mixed number to a percent:

   a. Multiply the fraction or mixed number by 100.

   b. Reduce, if possible.

   c. Affix a % sign.

   Illustration: Change $\frac{1}{7}$ to a percent.

   SOLUTION: $\frac{1}{7} \times 100 = \frac{100}{7}$
   $= 14\frac{2}{7}$
   $\frac{1}{7} = 14\frac{2}{7}\%$

   Answer: $14\frac{2}{7}\%$

   Illustration: Change $4\frac{2}{3}$ to a percent.

   SOLUTION: $4\frac{2}{3} \times 100 = \frac{14}{3} \times 100 = \frac{1400}{3}$
   $= 466\frac{2}{3}$
   $4\frac{2}{3} = 466\frac{2}{3}\%$

   Answer: $466\frac{2}{3}\%$

4. To remove a % sign attached to a decimal, divide the decimal by 100. If necessary, the resulting decimal may then be changed to a fraction.

Illustration: Change .5% to a decimal and to a fraction.

SOLUTION: $.5\% = .5 \div 100 = .005$
$.005 = \frac{5}{1000} = \frac{1}{200}$

Answer: $.5\% = .005$
$.5\% = \frac{1}{200}$

5. To remove a % sign attached to a fraction or mixed number, divide the fraction or mixed number by 100, and reduce, if possible. If necessary, the resulting fraction may then be changed to a decimal.

Illustration: Change $\frac{3}{4}\%$ to a fraction and to a decimal.

SOLUTION: $\frac{3}{4}\% = \frac{3}{4} \div 100 = \frac{3}{4} \times \frac{1}{100}$
$= \frac{3}{400}$

$\frac{3}{400} = 400 \overline{)3.0000}$ = .0075

Answer: $\frac{3}{4}\% = \frac{3}{400}$
$\frac{3}{4}\% = .0075$

6. To remove a % sign attached to a decimal that includes a fraction, divide the decimal by 100. If necessary, the resulting number may then be changed to a fraction.

Illustration: Change $.5\frac{1}{3}\%$ to a fraction.

SOLUTION: $.5\frac{1}{3}\% = .005\frac{1}{3}$
$= \frac{5\frac{1}{3}}{1000}$
$= 5\frac{1}{3} \div 1000$
$= \frac{16}{3} \times \frac{1}{1000}$
$= \frac{16}{3000}$
$= \frac{2}{375}$

Answer: $.5\frac{1}{3}\% = \frac{2}{375}$

7. Some fraction-percent equivalents are used so frequently that it is helpful to be familiar with them.

$\frac{1}{25} = 4\%$	$\frac{1}{5} = 20\%$
$\frac{1}{20} = 5\%$	$\frac{1}{4} = 25\%$
$\frac{1}{12} = 8\frac{1}{3}\%$	$\frac{1}{3} = 33\frac{1}{3}\%$
$\frac{1}{10} = 10\%$	$\frac{1}{2} = 50\%$
$\frac{1}{8} = 12\frac{1}{2}\%$	$\frac{2}{3} = 66\frac{2}{3}\%$
$\frac{1}{6} = 16\frac{2}{3}\%$	$\frac{3}{4} = 75\%$

Solving Percent Problems

8. Most percent problems involve three quantities:
   - The rate, R, which is followed by a % sign.
   - The base, B, which follows the word "of."
   - The amount or percentage, P, which usually follows the word "is."

a. If the rate (R) and the base (B) are known, then the percentage (P) = R × B.

Illustration: Find 15% of 50.

SOLUTION:

$$\begin{aligned} \text{Rate} &= 15\% \\ \text{Base} &= 50 \\ P &= R \times B \\ P &= 15\% \times 50 \\ &= .15 \times 50 \\ &= 7.5 \end{aligned}$$

Answer: 15% of 50 is 7.5.

b. If the rate (R) and the percentage (P) are known, then the base (B) = $\frac{P}{R}$.

Illustration: 7% of what number is 35?

SOLUTION:

$$\begin{aligned} \text{Rate} &= 7\% \\ \text{Percentage} &= 35 \\ B &= \frac{P}{R} \\ B &= \frac{35}{7\%} \\ &= 35 \div .07 \\ &= 500 \end{aligned}$$

Answer: 7% of 500 is 35.

c. If the percentage (P) and the base (B) are known, the rate (R) = $\frac{P}{B}$.

Illustration: There are 96 men in a group of 150 people. What percent of the group are men?

SOLUTION:

$$\begin{aligned} \text{Base} &= 150 \\ \text{Percentage (amount)} &= 96 \\ \text{Rate} &= \tfrac{96}{150} \\ &= .64 \\ &= 64\% \end{aligned}$$

Answer: 64% of the group are men.

Illustration: In a tank holding 20 gallons of solution, 1 gallon is alcohol. What is the strength of the solution in percent?

SOLUTION:

$$\begin{aligned} \text{Percentage (amount)} &= 1 \text{ gallon} \\ \text{Base} &= 20 \text{ gallons} \\ \text{Rate} &= \tfrac{1}{20} \\ &= .05 \\ &= 5\% \end{aligned}$$

Answer: The solution is 5% alcohol.

9. In a percent problem, the whole is 100%.

   Example: If a problem involves 10% of a quantity, the rest of the quantity is 90%.

   Example: If a quantity has been increased by 5%, the new amount is 105% of the original quantity.

   Example: If a quantity has been decreased by 15%, the new amount is 85% of the original quantity.

Practice Problems Involving Percents

1. 10% written as a decimal is
   (A) 1.0
   (B) 0.01
   (C) 0.001
   (D) 0.1

2. What is 5.37% in fraction form?
   (A) $\frac{537}{10.000}$
   (B) $5\frac{37}{10.000}$
   (C) $\frac{537}{1000}$
   (D) $5\frac{37}{100}$

3. What percent of $\frac{5}{6}$ is $\frac{3}{4}$?
   (A) 75%
   (B) 60%
   (C) 80%
   (D) 90%

4. What percent is 14 of 24?
   (A) $62\frac{1}{4}$%
   (B) $58\frac{1}{3}$%
   (C) $41\frac{2}{3}$%
   (D) $33\frac{3}{5}$%

5. 200% of 800 equals
   (A) 2500
   (B) 16
   (C) 1600
   (D) 4

6. If John must have a mark of 80% to pass a test of 35 items, the number of items he may miss and still pass the test is
   (A) 7
   (B) 8
   (C) 11
   (D) 28

7. The regular price of a TV set that sold for $118.80 at a 20% reduction sale is
   (A) $148.50
   (B) $142.60
   (C) $138.84
   (D) $ 95.04

8. A circle graph of a budget shows the expenditure of 26.2% for housing, 28.4% for food, 12% for clothing, 12.7% for taxes, and the balance for miscellaneous items. The percent for miscellaneous items is
   (A) 31.5
   (B) 79.3
   (C) 20.7
   (D) 68.5

9. Two dozen shuttlecocks and four badminton rackets are to be purchased for a playground. The shuttlecocks are priced at $.35 each and the rackets at $2.75 each. The playground receives a discount of 30% from these prices. The total cost of this equipment is
   (A) $ 7.29
   (B) $11.43
   (C) $13.58
   (D) $18.60

10. A piece of wood weighing 10 ounces is found to have a weight of 8 ounces after drying. The moisture content was
    (A) 25%
    (B) $33\frac{1}{3}$%
    (C) 20%
    (D) 40%

11. A bag contains 800 coins. Of these, 10 percent are dimes, 30 percent are nickels, and the rest are quarters. The amount of money in the bag is
    (A) less than $150
    (B) between $150 and $300
    (C) between $301 and $450
    (D) more than $450

12. Six quarts of a 20% solution of alcohol in water are mixed with 4 quarts of a 60% solution of alcohol in water. The alcoholic strength of the mixture is
    (A) 80% (C) 36%
    (B) 40% (D) 72%

13. A man insures 80% of his property and pays a $2\frac{1}{2}$% premium amounting to $348. What is the total value of his property?
    (A) $17,000 (C) $18,400
    (B) $18,000 (D) $17,400

14. A clerk divided his 35-hour work week as follows: $\frac{1}{5}$ of his time was spent in sorting mail; $\frac{1}{2}$ of his time in filing letters; and $\frac{1}{7}$ of his time in reception work. The rest of his time was devoted to messenger work. The percent of time spent on messenger work by the clerk during the week was most nearly
    (A) 6% (C) 14%
    (B) 10% (D) 16%

15. In a school in which 40% of the enrolled students are boys, 80% of the boys are present on a certain day. If 1152 boys are present, the total school enrollment is
    (A) 1440 (C) 3600
    (B) 2880 (D) 5400

Percent Problems — Correct Answers

1. **(D)**	6. **(A)**	11. **(A)**
2. **(A)**	7. **(A)**	12. **(C)**
3. **(D)**	8. **(C)**	13. **(D)**
4. **(B)**	9. **(C)**	14. **(D)**
5. **(C)**	10. **(C)**	15. **(C)**

Problem Solutions — Percents

1. 10% = .10 = .1

   Answer: **(D)** 0.1

2. $5.37\% = .0537 = \frac{537}{10,000}$

   Answer: **(A)** $\frac{537}{10,000}$

3. Base (number followed by "of") = $\frac{5}{6}$
   Percentage (number followed by "is") = $\frac{3}{4}$

   $$\text{Rate} = \frac{\text{Percentage}}{\text{Base}}$$
   $$= \text{Percentage} \div \text{Base}$$
   $$\text{Rate} = \frac{3}{4} \div \frac{5}{6}$$
   $$= \frac{3}{\not{4}_{2}} \times \frac{\not{6}^{3}}{5}$$
   $$= \frac{9}{10}$$
   $$\frac{9}{10} = .9 = 90\%$$

   Answer: **(D)** 90%

4. Base (number followed by "of") = 24
   Percentage (number followed by "is") = 14

   Rate = Percentage ÷ Base
   Rate = 14 ÷ 24
   = $.58\frac{1}{3}$
   = $58\frac{1}{3}$%

   Answer: **(B)** $58\frac{1}{3}$%

5. 200% of 800 = 2.00 × 800
   = 1600

Answer: **(C)** 1600

6. He must answer 80% of 35 correctly. Therefore, he may miss 20% of 35.

20% of 35 = .20 × 35
   = 7

Answer: **(A)** 7

7. Since $118.80 represents a 20% reduction, $118.80 = 80% of the regular price.

$$\text{Regular price} = \frac{\$118.80}{80\%}$$
   = $118.80 ÷ .80
   = $148.50

Answer: **(A)** $148.50

8. All the items in a circle graph total 100%. Add the figures given for housing, food, clothing, and taxes:

```
  26.2%
  28.4%
  12  %
+ 12.7%
-------
  79.3%
```

Subtract this total from 100% to find the percent for miscellaneous items:

```
  100.0%
-  79.3%
--------
   20.7%
```

Answer: **(C)** 20.7%

9. Price of shuttlecocks = 24 × $.35 = $ 8.40
Price of rackets = 4 × $2.75 = $11.00
Total price = $19.40

Discount is 30%, and 100% − 30% = 70%

Actual cost = 70% of 19.40
   = .70 × 19.40
   = 13.58

Answer: **(C)** $13.58

10. Subtract weight of wood after drying from original weight of wood to find amount of moisture in wood:

```
  10
-  8
```
2 ounces of moisture in wood

$$\text{Moisture content} = \frac{2 \text{ ounces}}{10 \text{ ounces}} = .2 = 20\%$$

Answer: **(C)** 20%

11. Find the number of each kind of coin:

10% of 800 = .10 × 800 = 80 dimes
30% of 800 = .30 × 800 = 240 nickels
60% of 800 = .60 × 800 = 480 quarters

Find the value of the coins:

80 dimes = 80 × .10 = $ 8.00
240 nickels = 240 × .05 = 12.00
480 quarters = 480 × .25 = 120.00
Total $140.00

Answer: **(A)** less than $150

12. First solution contains 20% of 6 quarts of alcohol.

Alcohol content = .20 × 6
   = 1.2 quarts

Second solution contains 60% of 4 quarts of alcohol.

Alcohol content = .60 × 4
   = 2.4 quarts

Mixture contains: 1.2 + 2.4 = 3.6 quarts alcohol
6 + 4 = 10 quarts liquid

$$\text{Alcoholic strength of mixture} = \frac{3.6}{10} = 36\%$$

Answer: **(C)** 36%

13. $2\frac{1}{2}$% of insured value = $348

$$\text{Insured value} = \frac{348}{2\frac{1}{2}\%}$$
   = 348 ÷ .025
   = $13,920

$13,920 is 80% of total value

$$\text{Total value} = \frac{\$13,920}{80\%}$$
   = $13,920 ÷ .80
   = $17,400

Answer: **(D)** $17,400

14. $\frac{1}{5} \times 35 =$ 7 hr sorting mail
$\frac{1}{2} \times 35 =$ $17\frac{1}{2}$ hr filing
$\frac{1}{7} \times 35 =$ 5 hr reception
$29\frac{1}{2}$ hr accounted for

$35 - 29\frac{1}{2}$ hr left for messenger work

% spent on messenger work:

$$= \frac{5\frac{1}{2}}{35}$$
$$= 5\frac{1}{2} \div 35$$
$$= \frac{11}{2} \times \frac{1}{35}$$
$$= \frac{11}{70}$$
$$= .15\frac{5}{7}$$

Answer: **(D)** most nearly 16% = $15\frac{5}{7}$

15. 80% of the boys = 1152

$$\text{Number of boys} = \frac{1152}{80\%}$$
$$= 1152 \div .80$$
$$= 1440$$

40% of students = 1440

$$\text{Total number of students} = \frac{1440}{40\%}$$
$$= 1440 \div .40$$
$$= 3600$$

Answer: **(C)** 3600

SHORTCUTS IN MULTIPLICATION AND DIVISION

There are several shortcuts for simplifying multiplication and division. Following the description of each shortcut, practice problems are provided.

Dropping Final Zeros

1. a. A zero in a whole number is considered a "final zero" if it appears in the units column or if all columns to its right are filled with zeros. A final zero may be omitted in certain kinds of problems.

   b. In decimal numbers a zero appearing in the extreme right column may be dropped with no effect on the solution of a problem.

2. In multiplying whole numbers, the final zero(s) may be dropped during computation and simply transferred to the answer.

Examples:

```
  2310        129       1760
 × 150      × 210      × 205
  1155        129        880
  231         258       352
 346500      27090     360800
```

Practice Problems

Solve the following multiplication problems, dropping the final zeros during computation.

```
1.   230          3.   203
   ×  12             ×  14

2.   175          4.   621
   × 130             × 140
```

5.
```
  430
× 360
```

6.
```
  132
× 310
```

7.
```
  350
×  24
```

8.
```
  520
× 410
```

9.
```
  634
× 120
```

10.
```
  431
× 230
```

Solutions to Practice Problems

1.
```
  230
×  12
   46
  23
 2760
```

2.
```
  175
× 130
  525
 175
22750
```

3.
```
  203
×  14
  812
 203
 2842
```
(no final zeros)

4.
```
  621
× 140
 2484
 621
86940
```

5.
```
   430
 × 360
   258
  129
154800
```

6.
```
  132
× 310
  132
 396
40920
```

7.
```
  350
×  24
  140
  70
 8400
```

8.
```
   520
 × 410
    52
  208
213200
```

9.
```
  634
× 120
 1268
 634
76080
```

10.
```
  431
× 230
 1293
 862
99130
```

Multiplying Whole Numbers by Decimals

3. In multiplying a whole number by a decimal number, if there are one or more final zeros in the multiplicand, move the decimal point in the multiplier to the right the same number of places as there are final zeros in the multiplicand. Then cross out the final zero(s) in the multiplicand.

Examples:

$$\begin{array}{r} 27500 \\ \times \quad .15 \\ \hline \end{array} = \begin{array}{r} 275 \\ \times \ 15 \\ \hline \end{array}$$

$$\begin{array}{r} 1250 \\ \times \ .345 \\ \hline \end{array} = \begin{array}{r} 125 \\ \times \ 3.45 \\ \hline \end{array}$$

Practice Problems

Rewrite the following problems, dropping the final zeros and moving decimal points the appropriate number of spaces. Then compute the answers.

1. $\begin{array}{r} 2400 \\ \times \quad .02 \\ \hline \end{array}$

2. $\begin{array}{r} 620 \\ \times \ .04 \\ \hline \end{array}$

3. $\begin{array}{r} 800 \\ \times \ .005 \\ \hline \end{array}$

4. $\begin{array}{r} 600 \\ \times \ .002 \\ \hline \end{array}$

5. $\begin{array}{r} 340 \\ \times \ .08 \\ \hline \end{array}$

6. $\begin{array}{r} 480 \\ \times \quad .4 \\ \hline \end{array}$

7. $\begin{array}{r} 400 \\ \times \ .04 \\ \hline \end{array}$

8. $\begin{array}{r} 5300 \\ \times \quad .5 \\ \hline \end{array}$

9. $\begin{array}{r} 930 \\ \times \quad .3 \\ \hline \end{array}$

10. $\begin{array}{r} 9000 \\ \times \ .001 \\ \hline \end{array}$

Solutions to Practice Problems

The rewritten problems are shown, along with the answers.

1. $\begin{array}{r} 24 \\ \times \ 2 \\ \hline 48 \end{array}$

2. $\begin{array}{r} 62 \\ \times \ .4 \\ \hline 24.8 \end{array}$

3. $8 \times .5 = 4.0$

4. $6 \times .2 = 1.2$

5. $34 \times .8 = 27.2$

6. $48 \times 4 = 192$

7. $4 \times 4 = 16$

8. $53 \times 50 = 2650$

9. $93 \times 3 = 279$

10. $9 \times 1 = 9$

Dividing by Whole Numbers

4. a. When there are final zeros in the divisor but no final zeros in the dividend, move the decimal point in the dividend to the left as many places as there are final zeros in the divisor, then omit the final zeros.

Example: $2700.\overline{)37523.} = 27.\overline{)375.23}$

b. When there are fewer final zeros in the divisor than there are in the dividend, drop the same number of final zeros from the dividend as there are final zeros in the divisor.

Example: $250.\overline{)45300.} = 25.\overline{)4530.}$

c. When there are more final zeros in the divisor than there are in the dividend, move the decimal point in the dividend to the left as many places as there are final zeros in the divisor, then omit the final zeros.

Example: $2300.\overline{)690.} = 23.\overline{)6.9}$

d. When there are no final zeros in the divisor, no zeros can be dropped in the dividend.

Example: $23.\overline{)690.} = 23.\overline{)690.}$

Practice Problems

Rewrite the following problems, dropping the final zeros and moving the decimal points the appropriate number of places. Then compute the quotients.

1. $600.\overline{)72.}$
2. $310.\overline{)6200.}$
3. $7600\overline{)1520.}$
4. $46.\overline{)920.}$
5. $11.0\overline{)220.}$
6. $700.\overline{)84.}$

7. $90.\,\overline{)\,8100.}$

8. $8100.\,\overline{)\,1620.}$

9. $25.\,\overline{)\,5250.}$

10. $41.0\,\overline{)\,820.}$

11. $800.\,\overline{)\,96.}$

12. $650.\,\overline{)\,1300.}$

13. $5500.\,\overline{)\,110.}$

14. $36.\,\overline{)\,720.}$

15. $87.0\,\overline{)\,1740.}$

Rewritten Practice Problems

1. $6.\,\overline{)\,.72}$

2. $31.\,\overline{)\,620.}$

3. $76.\,\overline{)\,15.2}$

4. $46.\,\overline{)\,920.}$

5. $11.\,\overline{)\,220.}$

6. $7.\,\overline{)\,.84}$

7. $9.\,\overline{)\,810.}$

8. $81.\,\overline{)\,16.2}$

9. $25.\,\overline{)\,5250.}$

10. $41.\,\overline{)\,820.}$

11. $8.\,\overline{)\,.96}$

12. $65.\,\overline{)\,130.}$

13. $55.\,\overline{)\,1.1}$

14. $36.\,\overline{)\,720.}$

15. $87.\,\overline{)\,1740.}$

Solutions to Practice Problems

1. $\begin{array}{r} .12 \\ 6.\,\overline{)\,.72} \end{array}$

2. $\begin{array}{r} 20 \\ 31.\,\overline{)\,620.} \\ \underline{62} \\ 00 \end{array}$

3. $\begin{array}{r} .2 \\ 76.\,\overline{)\,15.2} \\ \underline{15\;2} \\ 0\;0 \end{array}$

4. $\begin{array}{r} 20 \\ 46.\,\overline{)\,920.} \\ \underline{92} \\ 00 \end{array}$

5. $\begin{array}{r} 20 \\ 11.\,\overline{)\,220.} \\ \underline{22} \\ 00 \end{array}$

6. $\begin{array}{r} .12 \\ 7.\,\overline{)\,.84} \end{array}$

7. $\begin{array}{r} 90 \\ 9.\,\overline{)\,810.} \end{array}$

8. $\begin{array}{r} .2 \\ 81.\,\overline{)\,16.2} \\ \underline{16\;2} \\ 0\;0 \end{array}$

9. $\begin{array}{r} 210 \\ 25.\,\overline{)\,5250.} \\ \underline{50} \\ 25 \\ \underline{25} \\ 00 \end{array}$

10. $\begin{array}{r} 20 \\ 41.\,\overline{)\,820.} \\ \underline{82} \\ 00 \end{array}$

11. $\begin{array}{r} .12 \\ 8.\,\overline{)\,.96} \end{array}$

12. $\begin{array}{r} 2 \\ 65.\,\overline{)\,130.} \\ \underline{130} \\ 00 \end{array}$

13. $\begin{array}{r} .02 \\ 55.\,\overline{)\,1.10} \\ \underline{1\;10} \\ 00 \end{array}$

14. $\begin{array}{r} 20 \\ 36.\,\overline{)\,720.} \\ \underline{72} \\ 00 \end{array}$

15. $\begin{array}{r} 20 \\ 87.\,\overline{)\,1740.} \\ \underline{174} \\ 00 \end{array}$

Division by Multiplication

5. Instead of dividing by a particular number, the same answer is obtained by multiplying by the equivalent multiplier.

6. To find the equivalent multiplier of a given divisor, divide 1 by the divisor.

   Example: The equivalent multiplier of $12\frac{1}{2}$ is $1 \div 12\frac{1}{2}$ or .08. The division problem $100 \div 12\frac{1}{2}$ may be more easily solved as the multiplication problem $100 \times .08$. The answer will be the same.

7. Common divisors and their equivalent multipliers are shown below:

Divisor	*Equivalent Multiplier*
$11\frac{1}{9}$	.09
$12\frac{1}{2}$	.08
$14\frac{2}{7}$	.07
$16\frac{2}{3}$	.06
20	.05
25	.04
$33\frac{1}{3}$	.03
50	.02

8. A divisor may be multiplied or divided by any power of 10, and the only change in its equivalent multiplier will be in the placement of the decimal point, as may be seen in the following table:

Divisor	*Equivalent Multiplier*
.025	40.
.25	4.
2.5	.4
25.	.04
250.	.004
2500.	.0004

Practice Problems

Rewrite and solve each of the following problems by using equivalent multipliers. Drop the final zeros where appropriate.

1. $100 \div 16\frac{2}{3} =$
2. $200 \div 25 =$
3. $300 \div 33\frac{1}{3} =$
4. $250 \div 50 =$
5. $80 \div 12\frac{1}{2} =$
6. $800 \div 14\frac{2}{7} =$

7. $620 \div 20 =$
8. $500 \div 11\frac{1}{9} =$
9. $420 \div 16\frac{2}{3} =$
10. $1200 \div 33\frac{1}{3} =$
11. $955 \div 50 =$
12. $300 \div 33\frac{1}{3} =$
13. $275 \div 12\frac{1}{2} =$
14. $625 \div 25 =$
15. $244 \div 20 =$
16. $350 \div 16\frac{2}{3} =$
17. $400 \div 33\frac{1}{3} =$
18. $375 \div 25 =$
19. $460 \div 20 =$
20. $250 \div 12\frac{1}{2} =$

Solutions to Practice Problems

The rewritten problems and their solutions appear below:

1. $100 \times .06 = 1 \times 6 = 6$
2. $200 \times .04 = 2 \times 4 = 8$
3. $300 \times .03 = 3 \times 3 = 9$
4. $250 \times .02 = 25 \times .2 = 5$
5. $80 \times .08 = 8 \times .8 = 6.4$
6. $800 \times .07 = 8 \times 7 = 56$
7. $620 \times .05 = 62 \times .5 = 31$
8. $500 \times .09 = 5 \times 9 = 45$
9. $420 \times .06 = 42 \times .6 = 25.2$
10. $1200 \times .03 = 12 \times 3 = 36$
11. $955 \times .02 = 19.1$
12. $300 \times .03 = 3 \times 3 = 9$
13. $275 \times .08 = 22$
14. $625 \times .04 = 25$
15. $244 \times .05 = 12.2$
16. $350 \times .06 = 35 \times .6 = 21$
17. $400 \times .03 = 4 \times 3 = 12$
18. $375 \times .04 = 15$
19. $460 \times .05 = 46 \times .5 = 23$
20. $250 \times .08 = 25 \times .8 = 20$

Multiplication by Division

9. Just as some division problems are made easier by changing them to equivalent multiplication problems, certain multiplication problems are made easier by changing them to equivalent division problems.

10. Instead of arriving at an answer by multiplying by a particular number, the same answer is obtained by dividing by the equivalent divisor.

11. To find the equivalent divisor of a given multiplier, divide 1 by the multiplier.

12. Common multipliers and their equivalent divisors are shown below:

Multiplier	*Equivalent Divisor*
$11\frac{1}{9}$	.09
$12\frac{1}{2}$	.08
$14\frac{2}{7}$	.07
$16\frac{2}{3}$	.06
20	.05
25	.04
$33\frac{1}{3}$	.03
50	.02

Notice that the multiplier-equivalent divisor pairs are the same as the divisor-equivalent multiplier pairs given earlier.

Practice Problems

Rewrite and solve each of the following problems by using division. Drop the final zeros where appropriate.

1. $77 \times 14\frac{2}{7} =$
2. $81 \times 11\frac{1}{9} =$
3. $475 \times 20 =$
4. $42 \times 50 =$
5. $36 \times 33\frac{1}{3} =$
6. $96 \times 12\frac{1}{2} =$
7. $126 \times 16\frac{2}{3} =$
8. $48 \times 25 =$
9. $33 \times 33\frac{1}{3} =$
10. $84 \times 14\frac{2}{7} =$
11. $99 \times 11\frac{1}{9} =$
12. $126 \times 33\frac{1}{3} =$
13. $168 \times 12\frac{1}{2} =$
14. $654 \times 16\frac{2}{3} =$
15. $154 \times 14\frac{2}{7} =$
16. $5250 \times 50 =$
17. $324 \times 25 =$
18. $625 \times 20 =$
19. $198 \times 11\frac{1}{9} =$
20. $224 \times 14\frac{2}{7} =$

Solutions to Practice Problems

The rewritten problems and their solutions appear below:

1. $.07\overline{)77.} = 7\overset{1100.}{\overline{)7700.}}$

2. $.09\overline{)81.} = 9\overset{900.}{\overline{)8100.}}$

3. $.05\overline{)475.} = 5\overset{9500.}{\overline{)47500.}}$

4. $.02\overline{)42.} = 2\overset{2100.}{\overline{)4200.}}$

5. $.03\overline{)36.} = 3\overset{1200.}{\overline{)3600.}}$

6. $.08\overline{)96.} = 8\overset{1200.}{\overline{)9600.}}$

7. $.06\overline{)126.} = 6\overset{2100.}{\overline{)12600.}}$

8. $.04\overline{)48.} = 4\overset{1200.}{\overline{)4800.}}$

9. $.03\overline{)33.} = 3\overset{1100.}{\overline{)3300.}}$

10. $.07\overline{)84.} = 7\overset{1200.}{\overline{)8400.}}$

11. $.09\overline{)99.} = 9\overset{1100.}{\overline{)9900.}}$

12. $.03\overline{)126.} = 3\overset{4200.}{\overline{)12600.}}$

13. $.08\overline{)168.} = 8\overset{2100.}{\overline{)16800.}}$

14. $.06\overline{)654.} = 6\overset{10900.}{\overline{)65400.}}$

15. $.07\overline{)154.} = 7\overset{2200.}{\overline{)15400.}}$

16. $.02\overline{)5250.} = 2\overset{262500.}{\overline{)525000.}}$

17. $.04\overline{)324.} = 4\overset{8100.}{\overline{)32400.}}$

18. $.05\overline{)625.} = 5\overset{12500.}{\overline{)62500.}}$

19. $.09\overline{)198.} = 9\overset{2200.}{\overline{)19800.}}$

20. $.07\overline{)224.} = 7\overset{3200.}{\overline{)22400.}}$

POWERS AND ROOTS

1. The numbers that are multiplied to give a product are called the **factors** of the product.

   Example: In $2 \times 3 = 6$, 2 and 3 are factors.

2. If the factors are the same, an **exponent** may be used to indicate the number of times the factor appears.

   Example: In $3 \times 3 = 3^2$, the number 3 appears as a factor twice, as is indicated by the exponent 2.

3. When a product is written in exponential form, the number the exponent refers to is called the **base**. The product itself is called the **power**.

   Example: In 2^5, the number 2 is the base and 5 is the exponent.
   $2^5 = 2 \times 2 \times 2 \times 2 \times 2 = 32$, so 32 is the power.

4. a. If the exponent used is 2, we say that the base has been **squared**, or raised to the second power.

   Example: 6^2 is read "six squared" or "six to the second power."

   b. If the exponent used is 3, we say that the base has been **cubed**, or raised to the third power.

   Example: 5^3 is read "five cubed" or "five to the third power."

   c. If the exponent is 4, we say that the base has been raised to the fourth power. If the exponent is 5, we say the base has been raised to the fifth power, etc.

   Example: 2^8 is read "two to the eighth power."

5. A number that is the product of a number squared is called a **perfect square**.

   Example: 25 is a perfect square because $25 = 5^2$.

6. a. If a number has exactly two equal factors, each factor is called the **square root** of the number.

   Example: $9 = 3 \times 3$; therefore, 3 is the square root of 9.

   b. The symbol $\sqrt{\ }$ is used to indicate square root.

   Example: $\sqrt{9} = 3$ means that the square root of 9 is 3, or $3 \times 3 = 9$.

7. The square root of the most common perfect squares may be found by using the following table, or by trial and error; that is, by finding the number that, when squared, yields the given perfect square.

Number	*Perfect Square*	*Number*	*Perfect Square*
1	1	10	100
2	4	11	121
3	9	12	144
4	16	13	169
5	25	14	196
6	36	15	225
7	49	20	400
8	64	25	625
9	81	30	900

Example: To find $\sqrt{81}$, note that 81 is the perfect square of 9, or $9^2 = 81$. Therefore, $\sqrt{81} = 9$.

8. To find the square root of a number that is not a perfect square, use the following method:

   a. Locate the decimal point.

   b. Mark off the digits in groups of two in both directions beginning at the decimal point.

   c. Mark the decimal point for the answer just above the decimal point of the number whose square root is to be taken.

   d. Find the largest perfect square contained in the left-hand group of two.

   e. Place its square root in the answer. Subtract the perfect square from the first digit or pair of digits.

   f. Bring down the next pair.

   g. Double the partial answer.

   h. Add a trial digit to the right of the doubled partial answer. Multiply this new number by the trial digit. Place the correct new digit in the answer.

   i. Subtract the product.

   j. Repeat steps f–i as often as necessary.

   You will notice that you get one digit in the answer for every group of two you marked off in the original number.

   Illustration: Find the square root of 138,384.

SOLUTION:

$$\begin{array}{rr} & 3. \\ & \sqrt{13'83'84.} \\ 3^2 = & 9 \\ & 4\ 83 \end{array}$$

$$\begin{array}{rr} & 3\ \ 7\ \ 2. \\ & \sqrt{13'83'84.} \\ 3^2 = & 9 \\ & 4\ 83 \\ 7 \times 67 = & 4\ 69 \\ & 14\ 84 \\ 2 \times 742 = & 14\ 84 \end{array}$$

The number must first be marked off in groups of two figures each, beginning at the decimal point, which, in the case of a whole number, is at the right. The number of figures in the root will be the same as the number of groups so obtained.

The largest square less than 13 is 9. $\sqrt{9} = 3$

Place its square root in the answer. Subtract the perfect square from the first digit or pair of digits. Bring down the next pair. To form our trial divisor, annex 0 to this root "3" (making 30) and multiply by 2.

$483 \div 60 = 8$. Multiplying the trial divisor 68 by 8, we obtain 544, which is too large. We then try multiplying 67 by 7. This is correct. Add the trial digit to the right of the doubled partial answer. Place the new digit in the answer. Subtract the product. Bring down the final group. Annex 0 to the new root 37 and multiply by 2 for the trial divisor:

$$2 \times 370 = 740$$
$$1484 \div 740 = 2$$

Place the 2 in the answer.

Answer: The square root of 138,384 is 372.

Illustration: Find the square root of 3 to the nearest hundredth.

SOLUTION:

$$
\begin{array}{rl}
 & \phantom{\sqrt{}}1.\ 7\ \ 3\ \ 2 \\
 & \sqrt{3.00'00'00} \\
1^2 = & 1 \\
20 & 2\ 00 \\
7 \times 27 = & 1\ 89 \\
340 & \quad 11\ 00 \\
3 \times 343 = & \quad 10\ 29 \\
3460 & \qquad 71\ 00 \\
2 \times 3462 = & \qquad 69\ 24
\end{array}
$$

Answer: The square root of 3 is 1.73 to the nearest hundredth.

9. To find the square root of a fraction, find the square root of its numerator and of its denominator.

Example: $\sqrt{\frac{4}{9}} = \dfrac{\sqrt{4}}{\sqrt{9}} = \frac{2}{3}$

10. a. If a number has exactly three equal factors, each factor is called the **cube root** of the number.

b. The symbol $\sqrt[3]{\ }$ is used to indicate the cube root.

Example: $8 = 2 \times 2 \times 2$; therefore, $\sqrt[3]{8} = 2$

Practice Problems Involving Powers and Roots

1. The square of 10 is
   (A) 1 (C) 5
   (B) 2 (D) 100

2. The cube of 9 is
   (A) 3 (C) 81
   (B) 27 (D) 729

3. The fourth power of 2 is
   (A) 2 (C) 8
   (B) 4 (D) 16

4. In exponential form, the product $7 \times 7 \times 7 \times 7 \times 7$ may be written
   (A) 5^7 (C) 2^7
   (B) 7^5 (D) 7^2

5. The value of 3^5 is
   (A) 243 (C) 35
   (B) 125 (D) 15

6. The square root of 1175, to the nearest whole number, is
   (A) 32 (C) 34
   (B) 33 (D) 35

7. Find $\sqrt{503}$ to the nearest tenth.
   (A) 22.4 (C) 22.6
   (B) 22.5 (D) 22.7

8. Find $\sqrt{\frac{1}{4}}$.
   (A) 2 (C) $\frac{1}{8}$
   (B) $\frac{1}{2}$ (D) $\frac{1}{16}$

9. Find $\sqrt[3]{64}$.
   (A) 3 (C) 8
   (B) 4 (D) 32

10. The sum of 2^2 and 2^3 is
    (A) 9 (C) 12
    (B) 10 (D) 32

Powers and Roots Problems — Correct Answers

1. **(D)**
2. **(D)**
3. **(D)**
4. **(B)**
5. **(A)**
6. **(C)**
7. **(A)**
8. **(B)**
9. **(B)**
10. **(C)**

Problem Solutions — Powers and Roots

1. $10^2 = 10 \times 10 = 100$

   Answer: **(D)** 100

2. $9^3 = 9 \times 9 \times 9$
   $= 81 \times 9$
   $= 729$

   Answer: **(D)** 729

3. $2^4 = 2 \times 2 \times 2 \times 2$
$= 4 \times 2 \times 2$
$= 8 \times 2$
$= 16$

Answer: **(D)** 16

4. $7 \times 7 \times 7 \times 7 \times 7 = 7^5$

Answer: **(B)** 7^5

5. $3^5 = 3 \times 3 \times 3 \times 3 \times 3$
$= 243$

Answer: **(A)** 243

6.

```
              3  4. 2 = 34 to the nearest
           √11'75.00    whole number
    3² =     9
            2 75
4 × 64 =    2 56
              19 00
2 × 682 =     13 64
               5 36
```

Answer: **(C)** 34

7.

```
                2  2. 4  2 = 22.4 to the
              √5'03.00'00    nearest tenth
      2² =      4
               1 03
  2 × 42 =       84
                 19 00
 4 × 444 =       17 76
                  1 24 00
2 × 4482 =          89 64
                    34 36
```

Answer: **(A)** 22.4

8. $\sqrt{\frac{1}{4}} = \frac{\sqrt{1}}{\sqrt{4}} = \frac{1}{2}$

Answer: **(B)** $\frac{1}{2}$

9. Since $4 \times 4 \times 4 = 64$, $\sqrt[3]{64} = 4$

Answer: **(B)** 4

10. $2^2 + 2^3 = 4 + 8 = 12$

Answer: **(C)** 12

TABLE OF MEASURES

English Measures

Length

1 foot (ft or ′) = 12 inches (in or ″)
1 yard (yd) = 36 inches
1 yard = 3 feet
1 rod (rd) = $16\frac{1}{2}$ feet
1 mile (mi) = 5280 feet
1 mile = 1760 yards
1 mile = 320 rods

Liquid Measure

1 cup (c) = 8 fluid ounces (fl oz)
1 pint (pt) = 2 cups
1 pint = 4 gills (gi)
1 quart (qt) = 2 pints
1 gallon (gal) = 4 quarts
1 barrel (bl) = $31\frac{1}{2}$ gallons

Weight

1 pound (lb) = 16 ounces (oz)
1 hundredweight (cwt) = 100 pounds
1 ton (T) = 2000 pounds

Dry Measure

1 quart (qt) = 2 pints (pt)
1 peck (pk) = 8 quarts
1 bushel (bu) = 4 pecks

Area

1 square foot (ft^2) = 144 square inches (in^2)
1 square yard (yd^2) = 9 square feet

Volume

1 cubic foot (ft^3 or cu ft) = 1728 cubic inches
1 cubic yard (yd^3 or cu yd) = 27 cubic feet
1 gallon = 231 cubic inches

General Measures

Time

1 minute (min) = 60 seconds (sec)
1 hour (hr) = 60 minutes
1 day = 24 hours
1 week = 7 days
1 year = 52 weeks
1 calendar year = 365 days

Angles and Arcs

1 minute (′) = 60 seconds (″)
1 degree (°) = 60 minutes
1 circle = 360 degrees

Counting

1 dozen (doz) = 12 units
1 gross (gr) = 12 dozen
1 gross = 144 units

Table of English—Metric Conversions (Approximate)

English to Metric

1 inch = 2.54 centimeters
1 yard = .9 meters
1 mile = 1.6 kilometers
1 ounce = 28 grams
1 pound = 454 grams
1 fluid ounce = 30 milliliters
1 liquid quart = .95 liters

Metric to English

1 centimeter = .39 inches
1 meter = 1.1 yards
1 kilometer = .6 miles
1 kilogram = 2.2 pounds
1 liter = 1.06 liquid quart

*Table of Metric Conversions**

1 liter = 1000 cubic centimeters (cm^3)
1 milliliter = 1 cubic centimeter
1 liter of water weighs 1 kilogram
1 milliliter of water weighs 1 gram

*These conversions are exact only under specific conditions. If the conditions are not met, the conversions are approximate.

THE METRIC SYSTEM

LENGTH

Unit	Abbreviation	Number of Meters
myriameter	mym	10,000
kilometer	km	1,000
hectometer	hm	100
dekameter	dam	10
meter	m	1
decimeter	dm	0.1
centimeter	cm	0.01
millimeter	mm	0.001

AREA

Unit	Abbreviation	Number of Square Meters
square kilometer	sq km *or* km^2	1,000,000
hectare	ha	10,000
are	a	100
centare	ca	1
square centimeter	sq cm *or* cm^2	0.0001

VOLUME

Unit	Abbreviation	Number of Cubic Meters
dekastere	das	10
stere	s	1
decistere	ds	0.10
cubic centimeter	cu cm *or* cm^3 *or* cc	0.000001

CAPACITY

Unit	Abbreviation	Number of Liters
kiloliter	kl	1,000
hectoliter	hl	100
dekaliter	dal	10
liter	l	1
deciliter	dl	0.10
centiliter	cl	0.01
milliliter	ml	0.001

MASS AND WEIGHT

Unit	Abbreviation	Number of Grams
metric ton	MT *or* t	1,000,000
quintal	q	100,000
kilogram	kg	1,000
hectogram	hg	100
dekagram	dag	10
gram	g *or* gm	1
decigram	dg	0.10
centigram	cg	0.01
milligram	mg	0.001

DENOMINATE NUMBERS (MEASUREMENT)

1. A **denominate number** is a number that specifies a given measurement. The unit of measure is called the **denomination**.

   Example: 7 miles, 3 quarts, and 5 grams are denominate numbers.

2. a. The English system of measurement uses such denominations as pints, ounces, pounds, and feet.

   b. The metric system of measurement uses such denominations as grams, liters, and meters.

English System of Measurement

3. To convert from one unit of measure to another, find in the Table of Measures how many units of the smaller denomination equal one unit of the larger denomination. This number is called the **conversion number**.

4. To convert from one unit of measure to a smaller unit, multiply the given number of units by the conversion number.

   Illustration: Convert 7 yards to inches.

   SOLUTION: 1 yard = 36 inches (conversion number)
   7 yards = 7 × 36 inches
   = 252 inches

   Answer: 252 in

   Illustration: Convert 2 hours 12 minutes to minutes.

   SOLUTION: 1 hour = 60 minutes (conversion number)
   2 hr 12 min = 2 hr + 12 min
   2 hr = 2 × 60 min = 120 min
   2 hr 12 min = 120 min + 12 min
   = 132 min

   Answer: 132 min

5. To convert from one unit of measure to a larger unit:

   a. Divide the given number of units by the conversion number.

   Illustration: Convert 48 inches to feet.

   SOLUTION: 1 foot = 12 inches (conversion number)

   48 in ÷ 12 = 4 ft

   Answer: 4 ft

   b. If there is a remainder it is expressed in terms of the smaller unit of measure.

   Illustration: Convert 35 ounces to pounds and ounces.

   SOLUTION: 1 pound = 16 ounces (conversion number)

$$35 \text{ oz} \div 16 = 16 \overline{)\,35 \text{ oz}} \quad \begin{array}{r} 2 \text{ lb} \\ \underline{32} \\ 3 \text{ oz} \end{array}$$

   = 2 lb 3 oz

   Answer: 2 lb 3 oz

6. To add denominate numbers, arrange them in columns by common unit, then add each column. If necessary, simplify the answer, starting with the smallest unit.

   Illustration: Add 1 yd 2 ft 8 in, 2 yd 2 ft 10 in, and 3 yd 1 ft 9 in.

   SOLUTION:

```
    1 yd 2 ft  8 in
    2 yd 2 ft 10 in
  + 3 yd 1 ft  9 in
  -----------------
    6 yd 5 ft 27 in
  = 6 yd 7 ft  3 in   (since 27 in = 2 ft 3 in)
  = 8 yd 1 ft  3 in   (since 7 ft = 2 yd 1 ft)
```

   Answer: 8 yd 1 ft 3 in

7. To subtract denominate numbers, arrange them in columns by common unit,then subtract each column starting with the smallest unit. If necessary, borrow to increase the number of a particular unit.

   Illustration: Subtract 2 gal 3 qt from 7 gal 1 qt.

   SOLUTION:

```
    7 gal 1 qt =   6 gal 5 qt
  - 2 gal 3 qt = - 2 gal 3 qt
  ---------------------------
                   4 gal 2 qt
```

   Note that 1 gal was borrowed from 7 gal.

   1 gal = 4 qt

   Therefore, 7 gal 1 qt = 6 gal 5 qt

   Answer: 4 gal 2 qt

8. To multiply a denominate number by a given number:

   a. If the denominate number contains only one unit, multiply the numbers and write the unit.

   Example: 3 oz × 4 = 12 oz

b. If the denominate number contains more than one unit of measurement, multiply the number of each unit by the given number and simplify the answer, if necessary.

Illustration: Multiply 4 yd 2 ft 8 in by 2.

SOLUTION:

```
     4 yd  2 ft   8 in
×                    2
  --------------------
     8 yd  4 ft  16 in
   = 8 yd  5 ft   4 in  (since 16 in = 1 ft 4 in)
   = 9 yd  2 ft   4 in  (since 5 ft = 1 yd 2 ft)
```

Answer: 9 yd 2 ft 4 in

9. To divide a denominate number by a given number, convert all units to the smallest unit, then divide. Simplify the answer, if necessary.

Illustration: Divide 5 lb 12 oz by 4.

SOLUTION:

```
       1 lb = 16 oz, therefore
5 lb 12 oz = 92 oz
92 oz ÷ 4 = 23 oz
          = 1 lb 7 oz
```

Answer: 1 lb 7 oz

10. Alternate method of division:

a. Divide the number of the largest unit by the given number.

b. Convert any remainder to the next largest unit.

c. Divide the total number of that unit by the given number.

d. Again convert any remainder to the next unit and divide.

e. Repeat until no units remain.

Illustration: Divide 9 hr 21 min 40 sec by 4.

SOLUTION:

```
     2 hr    20 min      25 sec
4 ) 9 hr    21 min      40 sec
    8 hr
    ----
    1 hr = 60 min
           ------
           81 min
           80 min
           ------
           1 min =      60 sec
                       -------
                       100 sec
                       100 sec
                       -------
                         0 sec
```

Answer: 2 hr 20 min 25 sec

Metric Measurement

11. The basic units of the metric system are the meter (m), which is used for length; the gram (g), which is used for weight; and the liter (*l*), which is used for capacity, or volume.

12. The prefixes that are used with the basic units, and their meanings, are:

Prefix	*Abbreviation*	*Meaning*
micro	**m**	one millionth of (.000001)
milli	m	one thousandth of (.001)
centi	c	one hundredth of (.01)
deci	d	one tenth of (.1)
deka	da or dk	ten times (10)
hecto	h	one hundred times (100)
kilo	k	one thousand times (1000)
mega	M	one million times (1,000,000)

13. To convert *to* a basic metric unit from a prefixed metric unit, multiply by the number indicated in the prefix.

Example: Convert 72 millimeters to meters.

$$72 \text{ millimeters} = 72 \times .001 \text{ meters} = .072 \text{ meters}$$

Example: Convert 4 kiloliters to liters.

$$4 \text{ kiloliters} = 4 \times 1000 \text{ liters} = 4000 \text{ liters}$$

14. To convert *from* a basic unit to a prefixed unit, divide by the number indicated in the prefix.

Example: Convert 300 liters to hectoliters.

$$300 \text{ liters} = 300 \div 100 \text{ hectoliters} = 3 \text{ hectoliters}$$

Example: Convert 4.5 meters to decimeters.

$$4.5 \text{ meters} = 4.5 \div .1 \text{ decimeters} = 45 \text{ decimeters}$$

15. To convert from any prefixed metric unit to another prefixed unit, first convert to a basic unit, then convert the basic unit to the desired unit.

Illustration: Convert 420 decigrams to kilograms.

SOLUTION: $420 \text{ dg} = 420 \times .1 \text{ g} = 42 \text{ g}$

$42 \text{ g} = 42 \div 1000 \text{ kg} = .042 \text{ kg}$

Answer: .042 kg

16. To add, subtract, multiply, or divide using metric measurement, first convert all units to the same unit, then perform the desired operation.

Illustration: Subtract 1200 g from 2.5 kg.

SOLUTION:

$$\begin{array}{rcr} 2.5 \text{ kg} & = & 2500 \text{ g} \\ -\ 1200 \text{ g} & = & -\ 1200 \text{ g} \\ \hline & & 1300 \text{ g} \end{array}$$

Answer: 1300 g or 1.3 kg

17. To convert from a metric measure to an English measure, or the reverse:

a. In the Table of English–Metric Conversions, find how many units of the desired measure are equal to one unit of the given measure.

b. Multiply the given number by the number found in the table.

Illustration: Find the number of pounds in 4 kilograms.

SOLUTION: From the table, 1 kg = 2.2 lb.
4 kg = 4 × 2.2 lb
= 8.8 lb

Answer: 8.8 lb

Illustration: Find the number of meters in 5 yards.

SOLUTION: 1 yd = .9 m
5 yd = 5 × .9 m
= 4.5 m

Answer: 4.5 m

Temperature Measurement

18. The temperature measurement currently used in the United States is the degree Fahrenheit (°F). The metric measurement for temperature is the degree Celsius (°C), also called degree Centigrade.

19. Degrees Celsius may be converted to degrees Fahrenheit by the formula:

$$°F = \frac{9}{5}°C + 32°$$

Illustration: Water boils at 100°C. Convert this to °F.

SOLUTION: $°F = \frac{9}{\cancel{5}_1} \times \cancel{100}^{20}° + 32°$
$= 180° + 32°$
$= 212°$

Answer: 100°C = 212°F

20. Degrees Fahrenheit may be converted to degrees Celsius by the formula:

$$°C = \frac{5}{9}(°F - 32°)$$

In using this formula, perform the subtraction in the parentheses first, then multiply by $\frac{5}{9}$.

Illustration: If normal body temperature is 98.6°F, what is it on the Celsius scale?

SOLUTION:
$$\begin{aligned} °C &= \tfrac{5}{9}(98.6° - 32°) \\ &= \tfrac{5}{9} \times 66.6° \\ &= \tfrac{333}{9}° \\ &= 37° \end{aligned}$$

Answer: Normal body temperature = 37°C.

Practice Problems Involving Measurement

1. A carpenter needs boards for 4 shelves, each 2′9″ long. How many feet of board should he buy?
(A) 11 (C) 13
(B) $11\frac{1}{6}$ (D) $15\frac{1}{2}$

2. The number of half-pints in 19 gallons of milk is
(A) 76 (C) 304
(B) 152 (D) 608

3. The product of 8 ft 7 in multiplied by 8 is
(A) 69 ft 6 in (C) $68\frac{2}{3}$ ft
(B) 68.8 ft (D) 68 ft 2 in

4. $\frac{1}{3}$ of 7 yards is
(A) 2 yd (C) $3\frac{1}{2}$ yd
(B) 4 ft (D) 7 ft

5. Six gross of special drawing pencils were purchased for use in an office. If the pencils were used at the rate of 24 a week, the maximum number of weeks that the six gross of pencils would last is
(A) 6 weeks (C) 24 weeks
(B) 12 weeks (D) 36 weeks

6. If 7 ft 9 in is cut from a piece of wood that is 9 ft 6 in, the piece left is
(A) 1 ft 9 in (C) 2 ft 2 in
(B) 1 ft 10 in (D) 2 ft 5 in

7. Take 3 hours 49 minutes from 5 hours 13 minutes.
(A) 1 hr 5 min (C) 1 hr 18 min
(B) 1 hr 10 min (D) 1 hr 24 min

8. A piece of wood 35 feet 6 inches long was used to make 4 shelves of equal lengths. The length of each shelf was
(A) 8.9 in (C) 8 ft $9\frac{1}{2}$ in
(B) 8 ft 9 in (D) 8 ft $10\frac{1}{2}$ in

9. The number of yards equal to 126 inches is
(A) 3.5 (C) 1260
(B) 10.5 (D) 1512

10. If there are 231 cubic inches in one gallon, the number of cubic inches in 3 pints is closest to which one of the following?
(A) 24 (C) 57
(B) 29 (D) 87

11. The sum of 5 feet $2\frac{3}{4}$ inches, 8 feet $\frac{1}{2}$ inch, and $12\frac{1}{2}$ inches is
(A) 14 ft $3\frac{3}{4}$ in (C) 14 ft $9\frac{1}{4}$ in
(B) 14 ft $5\frac{3}{4}$ in (D) 15 ft $\frac{1}{2}$ in

12. Add 5 hr 13 min, 3 hr 49 min, and 14 min. The sum is
(A) 8 hr 16 min (C) 9 hr 76 min
(B) 9 hr 16 min (D) 8 hr 6 min

13. Assuming that 2.54 centimeters = 1 inch, a metal rod that measures $1\frac{1}{2}$ feet would most nearly equal which one of the following?
(A) 380 cm (C) 30 cm
(B) 46 cm (D) 18 cm

14. A micromillimeter is defined as one millionth of a millimeter. A length of 17 micromillimeters may be represented as
(A) .00017 mm (C) .000017 mm
(B) .0000017 mm (D) .00000017 mm

15. How many liters are equal to 4200 m*l*?
(A) .42 (C) 420
(B) 4.2 (D) 420,000

16. Add 26 dg, .4 kg, 5 g, and 184 cg.
(A) 215.40 g (C) 409.44 g
(B) 319.34 g (D) 849.00 g

17. Four full bottles of equal size contain a total of 1.28 liters of cleaning solution. How many milliliters are in each bottle?
(A) 3.20 (C) 320
(B) 5.12 (D) 512

18. How many liters of water can be held in a 5-gallon jug? (See Conversion Table.)
(A) 19 (C) 40
(B) 38 (D) 50

19. To the nearest degree, what is a temperature of 12°C equal to on the Fahrenheit scale?
(A) 19° (C) 57°
(B) 54° (D) 79°

20. A company requires that the temperature in its offices be kept at 68°F. What is this in °C?
(A) 10° (C) 20°
(B) 15° (D) 25°

Measurement Problems — Correct Answers

1. **(A)**
2. **(C)**
3. **(C)**
4. **(D)**
5. **(D)**
6. **(A)**
7. **(D)**
8. **(D)**
9. **(A)**
10. **(D)**
11. **(A)**
12. **(B)**
13. **(B)**
14. **(C)**
15. **(B)**
16. **(C)**
17. **(C)**
18. **(A)**
19. **(B)**
20. **(C)**

Problem Solutions — Measurement

1.
$$\begin{array}{r} 2 \text{ ft } \ 9 \text{ in} \\ \times \qquad \quad 4 \\ \hline 8 \text{ ft } 36 \text{ in} \end{array} = 11 \text{ ft}$$

Answer: **(A)** 11

2. Find the number of half-pints in 1 gallon:

1 gal = 4 qts
4 qts = 4 × 2 pts = 8 pts
8 pts = 8 × 2 = 16 half-pints

Multiply to find the number of half-pints in 19 gallons:

19 gal = 19 × 16 half-pints
= 304 half-pints

Answer: **(C)** 304

3. 8 ft 7 in
× 8
64 ft 56 in = 68 ft 8 in
(since 56 in = 4 ft 8 in)

$8 \text{ in} = \frac{8}{12} \text{ ft} = \frac{2}{3} \text{ ft}$

$68 \text{ ft } 8 \text{ in} = 68\frac{2}{3} \text{ ft}$

Answer: **(C)** $68\frac{2}{3}$ ft

4. $\frac{1}{3} \times 7 \text{ yd} = 2\frac{1}{3} \text{ yd}$
= 2 yd 1 ft
= 2 × 3 ft + 1 ft
= 7 ft

Answer: **(D)** 7 ft

5. Find the number of units in 6 gross:
1 gross = 144 units
6 gross = 6 × 144 units
= 864 units

Divide units by rate of use:
864 ÷ 24 = 36

Answer: **(D)** 36 weeks

6. 9 ft 6 in = 8 ft 18 in
− 7 ft 9 in = − 7 ft 9 in
1 ft 9 in

Answer: **(A)** 1 ft 9 in

7. 5 hours 13 minutes = 4 hours 73 minutes
− 3 hours 49 minutes = − 3 hours 49 minutes
1 hour 24 minutes

Answer: **(D)** 1 hr 24 min

8. 4) 35 feet 6 inches
Quotient: 8 feet 10 inches + $\frac{2}{4}$ inches = 8 ft $10\frac{1}{2}$ in
32 feet
3 feet = 36 inches
42 inches
40 inches
2 inches

Answer: **(D)** 8 ft $10\frac{1}{2}$ in

9. 1 yd = 36 in
126 ÷ 36 = 3.5

Answer: **(A)** 3.5

10. 1 gal = 4 qt = 8 pt
Therefore, 1 pt = 231 cubic inches ÷ 8
= 28.875 cubic inches
3 pts = 3 × 28.875 cubic inches
= 86.625 cubic inches

Answer: **(D)** 87

11. 5 feet $2\frac{3}{4}$ inches
8 feet $\frac{1}{2}$ inches
+ $12\frac{1}{2}$ inches
13 feet $15\frac{3}{4}$ inches
= 14 feet $3\frac{3}{4}$ inches

Answer: **(A)** 14 feet $3\frac{3}{4}$ inches

12. 5 hr 13 min
3 hr 49 min
+ 14 min
8 hr 76 min
= 9 hr 16 min

Answer: **(B)** 9 hr 16 min

13. 1 foot = 12 inches
$1\frac{1}{2}$ feet = $1\frac{1}{2}$ × 12 inches = 18 inches
1 inch = 2.54 cm
Therefore,
18 inches = 18 × 2.54 cm
= 45.72 cm

Answer: **(B)** 46 cm

14. 1 micromillimeter = .000001 mm
17 micromillimeters = 17 × .000001 mm
= .000017 mm

Answer: **(C)** .000017 mm

15. 4200 m*l* = 4200 × .001 *l*
= 4.200 *l*

Answer: **(B)** 4.2

16. Convert all of the units to grams:
26 dg = 26 × .1 g = 2.6 g
.4 kg = .4 × 1000 g = 400 g
5 g = 5 g
184 cg = 184 × .01 g = 1.84 g
409.44 g

Answer: **(C)** 409.44 g

17. $1.28 \text{ liters} \div 4 = .32 \text{ liters}$

$$.32 \text{ liters} = .32 \div .001 \text{ m}l$$
$$= 320 \text{ m}l$$

Answer: **(C)** 320

18. Find the number of liters in 1 gallon:

$$1 \text{ qt} = .95\ l$$
$$1 \text{ gal} = 4 \text{ qts}$$
$$1 \text{ gal} = 4 \times .95\ l = 3.8\ l$$

Multiply to find the number of liters in 5 gallons:

$$5 \text{ gal} = 5 \times 3.8\ l = 19\ l$$

Answer: **(A)** 19

19.

$$°F = \tfrac{9}{5} \times 12° + 32°$$
$$= \tfrac{108°}{5} + 32°$$
$$= 21.6° + 32°$$
$$= 53.6°$$

Answer: **(B)** 54°

20.

$$°C = \tfrac{5}{9}(68° - 32°)$$
$$= \tfrac{\cancel{5}}{\cancel{9}_1} \times \overset{4}{\cancel{36}°}$$
$$= 20°$$

Answer: **(C)** 20°

STATISTICS AND PROBABILITY

Statistics

1. The **averages** used in statistics include the **arithmetic mean**, the **median** and the **mode**.

2. a. The most commonly used average of a group of numbers is the **arithmetic mean**. It is found by adding the numbers given and then dividing this sum by the number of items being averaged.

Illustration: Find the arithmetic mean of 2, 8, 5, 9, 6, and 12.

SOLUTION: There are 6 numbers.

$$\text{Arithmetic mean} = \frac{2 + 8 + 5 + 9 + 6 + 12}{6} = \frac{42}{6} = 7$$

Answer: The arithmetic mean is 7.

b. If a problem calls for simply the "average" or the "mean," it is referring to the arithmetic mean.

3. If a group of numbers is arranged in order, the middle number is called the **median**. If there is no single middle number (this occurs when there is an even number of items), the median is found by computing the arithmetic mean of the two middle numbers.

Example: The median of 6, 8, 10, 12, and 14 is 10.

Example: The median of 6, 8, 10, 12, 14, and 16 is the arithmetic mean of 10 and 12.

$$\frac{10 + 12}{2} = \frac{22}{2} = 11.$$

4. The **mode** of a group of numbers is the number that appears most often.

Example: The mode of 10, 5, 7, 9, 12, 5, 10, 5 and 9 is 5.

5. To obtain the average of quantities that are weighted:

a. Set up a table listing the quantities, their respective weights, and their respective values.

b. Multiply the value of each quantity by its respective weight.

c. Add up these products.

d. Add up the weights.

e. Divide the sum of the products by the sum of the weights.

Illustration: Assume that the weights for the following subjects are: English 3, History 2, Mathematics 2, Foreign Languages 2, and Art 1. What would be the average of a student whose marks are: English 80, History 85, Algebra 84, Spanish 82, and Art 90?

SOLUTION:

Subject	*Weight*	*Mark*
English	3	80
History	2	85
Algebra	2	84
Spanish	2	82
Art	1	90

English	$3 \times 80 = 240$
History	$2 \times 85 = 170$
Algebra	$2 \times 84 = 168$
Spanish	$2 \times 82 = 164$
Art	$1 \times 90 = 90$
	832

Sum of the weights: $3 + 2 + 2 + 2 + 1 = 10$

$832 \div 10 = 83.2$

Answer: Average = 83.2

Probability

6. The study of probability deals with predicting the outcome of chance events; that is, events in which one has no control over the results.

   Examples: Tossing a coin, rolling dice, and drawing concealed objects from a bag are chance events.

7. The probability of a particular outcome is equal to the number of ways that outcome can occur, divided by the total number of possible outcomes.

   Example: In tossing a coin, there are 2 possible outcomes: heads or tails. The probability that the coin will turn up heads is $1 \div 2$ or $\frac{1}{2}$.

   Example: If a bag contains 5 balls of which 3 are red, the probability of drawing a red ball is $\frac{3}{5}$. The probability of drawing a non-red ball is $\frac{2}{5}$.

8. a. If an event is certain, its probability is 1.

   Example: If a bag contains only red balls, the probability of drawing a red ball is 1.

   b. If an event is impossible, its probability is 0.

   Example: If a bag contains only red balls, the probability of drawing a green ball is 0.

9. Probability may be expressed in fractional, decimal, or percent form.

   Example: An event having a probability of $\frac{1}{2}$ is said to be 50% probable.

10. A probability determined by random sampling of a group of items is assumed to apply to other items in that group and in other similar groups.

    Illustration: A random sampling of 100 items produced in a factory shows that 7 are defective. How many items of the total production of 50,000 can be expected to be defective?

    SOLUTION: The probability of an item being defective is $\frac{7}{100}$, or 7%. Of the total production, 7% can be expected to be defective.

    $$7\% \times 50{,}000 = .07 \times 50{,}000 = 3500$$

    Answer: 3500 items

Practice Problems Involving Statistics and Probability

1. The arithmetic mean of 73.8, 92.2, 64.7, 43.8, 56.5, and 46.4 is
   (A) 60.6 (C) 64.48
   (B) 62.9 (D) 75.48

2. The median of the numbers 8, 5, 7, 5, 9, 9, 1, 8, 10, 5, and 10 is
   (A) 5 (C) 8
   (B) 7 (D) 9

3. The mode of the numbers 16, 15, 17, 12, 15, 15, 18, 19, and 18 is
   (A) 15 (C) 17
   (B) 16 (D) 18

4. A clerk filed 73 forms on Monday, 85 forms on Tuesday, 54 on Wednesday, 92 on Thursday, and 66 on Friday. What was the average number of forms filed per day?
   (A) 60 (C) 74
   (B) 72 (D) 92

5. The grades received on a test by twenty students were: 100, 55, 75, 80, 65, 65, 85, 90, 80, 45, 40, 50, 85, 85, 85, 80, 80, 70, 65, and 60. The average of these grades is
   (A) 70 (C) 77
   (B) 72 (D) 80

6. A buyer purchased 75 six-inch rulers costing 15¢ each, 100 one-foot rulers costing 30¢ each, and 50 one-yard rulers costing 72¢ each. What was the average price per ruler?
   (A) $26\frac{1}{8}$¢ (C) 39¢
   (B) $34\frac{1}{3}$¢ (D) 42¢

7. What is the average of a student who received 90 in English, 84 in Algebra, 75 in French, and 76 in Music, if the subjects have the following weights: English 4, Algebra 3, French 3, and Music 1?
   (A) 81 (C) 82
   (B) $81\frac{1}{2}$ (D) 83

Questions 8–11 refer to the following information:

A census shows that on a certain block the number of children in each family is 3, 4, 4, 0, 1, 2, 0, 2, and 2, respectively.

8. Find the average number of children per family.
   (A) 2 (C) 3
   (B) $2\frac{1}{2}$ (D) $3\frac{1}{2}$

9. Find the median number of children.
   (A) 1 (C) 3
   (B) 2 (D) 4

10. Find the mode of the number of children.
    (A) 0 (C) 2
    (B) 1 (D) 4

11. What is the probability that a family chosen at random on this block will have 4 children?
    (A) $\frac{4}{9}$ (C) $\frac{4}{7}$
    (B) $\frac{2}{9}$ (D) $\frac{2}{1}$

12. What is the probability that an even number will come up when a single die is thrown?
    (A) $\frac{1}{6}$ (C) $\frac{1}{2}$
    (B) $\frac{1}{3}$ (D) 1

13. A bag contains 3 black balls, 2 yellow balls, and 4 red balls. What is the probability of drawing a black ball?
    (A) $\frac{1}{2}$ (C) $\frac{2}{3}$
    (B) $\frac{1}{3}$ (D) $\frac{4}{9}$

14. In a group of 1000 adults, 682 are women. What is the probability that a person chosen at random from this group will be a man?
    (A) .318 (C) .5
    (B) .682 (D) 1

15. In a balloon factory, a random sampling of 100 balloons showed that 3 had pinholes in them. In a sampling of 2500 balloons, how many may be expected to have pinholes?
    (A) 30 (C) 100
    (B) 75 (D) 750

Statistics and Probability Problems — Correct Answers

1. **(B)**	6. **(B)**	11. **(B)**
2. **(C)**	7. **(D)**	12. **(C)**
3. **(A)**	8. **(A)**	13. **(B)**
4. **(C)**	9. **(B)**	14. **(A)**
5. **(B)**	10. **(C)**	15. **(B)**

Problem Solutions — Statistics and Probability

1. Find the sum of the values:

   $73.8 + 92.2 + 64.7 + 43.8 + 56.5 + 46.4 = 377.4$

   There are 6 values.

   $$\text{Arithmetic mean} = \frac{377.4}{6} = 62.9$$

   Answer: **(B)** 62.9

2. Arrange the numbers in order:

   1, 5, 5, 5, 7, 8, 8, 9, 9, 10, 10

   The middle number, or median, is 8.

   Answer: **(C)** 8

3. The mode is that number appearing most frequently. The number 15 appears three times.

   Answer: **(A)** 15

4. $\text{Average} = \frac{73 + 85 + 54 + 92 + 66}{5}$

$= \frac{370}{5}$

$= 74$

Answer: **(C)** 74

5. Sum of the grades = 1440.

$\frac{1440}{20} = 72$

Answer: **(B)** 72

6.

$75 \times 15¢ = 1125¢$
$100 \times 30¢ = 3000¢$
$\underline{50} \times 72¢ = \underline{3600¢}$
$225 \qquad 7725¢$

$\frac{7725¢}{225} = 34\frac{1}{3}¢$

Answer: **(B)** $34\frac{1}{3}$¢

7.

Subject	*Grade*	*Weight*
English	90	4
Algebra	84	3
French	75	3
Music	76	1

$(90 \times 4) + (84 \times 3) + (75 \times 3) + (76 \times 1)$

$360 + 252 + 225 + 76 = 913$

$\text{Weight} = 4 + 3 + 3 + 1 = 11$

$913 \div 11 = 83$ average

Answer: **(D)** 83

8. $\text{Average} = \frac{3 + 4 + 4 + 0 + 1 + 2 + 0 + 2 + 2}{9}$

$= \frac{18}{9}$

$= 2$

Answer: **(A)** 2

9. Arrange the numbers in order:

0, 0, 1, 2, 2, 2, 3, 4, 4

Of the 9 numbers, the fifth (middle) number is 2.

Answer: **(B)** 2

10. The number appearing most often is 2.

Answer: **(C)** 2

11. There are 9 families, 2 of which have 4 children. The probability is $\frac{2}{9}$.

Answer: **(B)** $\frac{2}{9}$

12. Of the 6 possible numbers, three are even (2, 4, and 6). The probability is $\frac{3}{6}$, or $\frac{1}{2}$.

Answer: **(C)** $\frac{1}{2}$

13. There are 9 balls in all. The probability of drawing a black ball is $\frac{3}{9}$, or $\frac{1}{3}$.

Answer: **(B)** $\frac{1}{3}$

14. If 682 people of the 1000 are women, $1000 - 682 = 318$ are men. The probability of choosing a man is $\frac{318}{1000} = .318$.

Answer: **(A)** .318

15. There is a probability of $\frac{3}{100} = 3\%$ that a balloon may have a pinhole.

$3\% \times 2500 = 75.00$

Answer: **(B)** 75

GRAPHS

1. **Graphs** illustrate comparisons and trends in statistical information. The most commonly used graphs are **bar graphs**, **line graphs**, **circle graphs**, and **pictographs**.

Bar Graphs

2. **Bar graphs** are used to compare various quantities. Each bar may represent a single quantity or may be divided to represent several quantities.

3. Bar graphs may have horizontal or vertical bars.
Illustration:

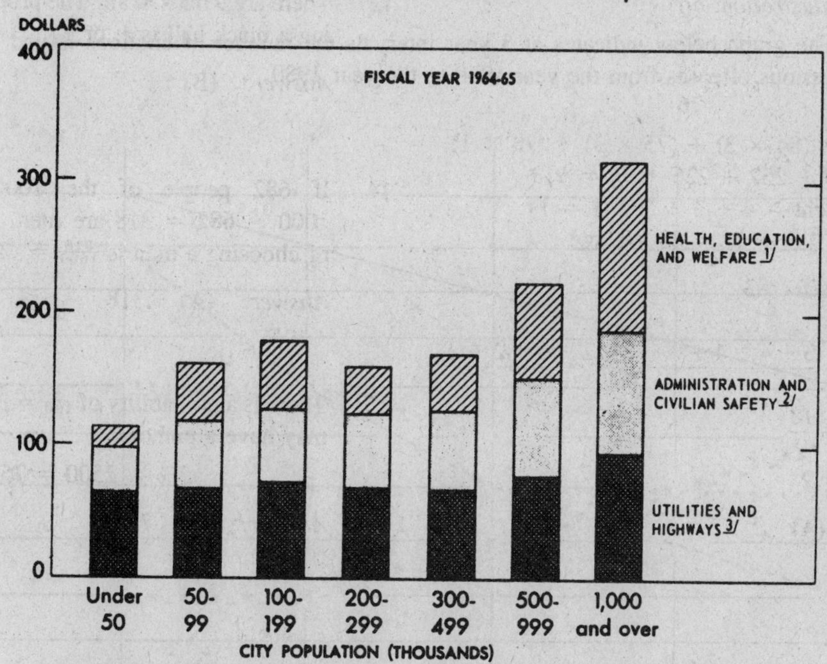

1/PUBLIC WELFARE, EDUCATION, HOSPITALS, HEALTH, LIBRARIES, AND HOUSING AND URBAN RENEWAL.
2/POLICE AND FIRE PROTECTION, FINANCIAL ADMINISTRATION, GENERAL CONTROL, GENERAL PUBLIC BUILDINGS, INTEREST ON GENERAL DEBT, AND OTHER.
3/HIGHWAYS, SEWERAGE, SANITATION, PARKS AND RECREATION, AND UTILITIES.
SOURCE: DEPARTMENT OF COMMERCE.

Question 1: What was the approximate municipal expenditure per capita in cities having populations of 200,000 to 299,000?

Answer: The middle bar of the seven shown represents cities having populations from 200,000 to 299,000. This bar reaches about halfway between 100 and 200. Therefore, the per capita expenditure was approximately $150.

Question 2: Which cities spent the most per capita on health, education, and welfare?

Answer: The bar for cities having populations of 1,000,000 and over has a larger striped section than the other bars. Therefore, those cities spent the most.

Question 3: Of the three categories of expenditures, which was least dependent on city size?

Answer: The expenditures for utilities and highways, the darkest part of each bar, varied least as city size increased.

Line Graphs

4. **Line graphs** are used to show trends, often over a period of time.

5. A line graph may include more than one line, with each line representing a different item.

Illustration:

The graph below indicates at 5 year intervals the number of citations issued for various offenses from the year 1960 to the year 1980.

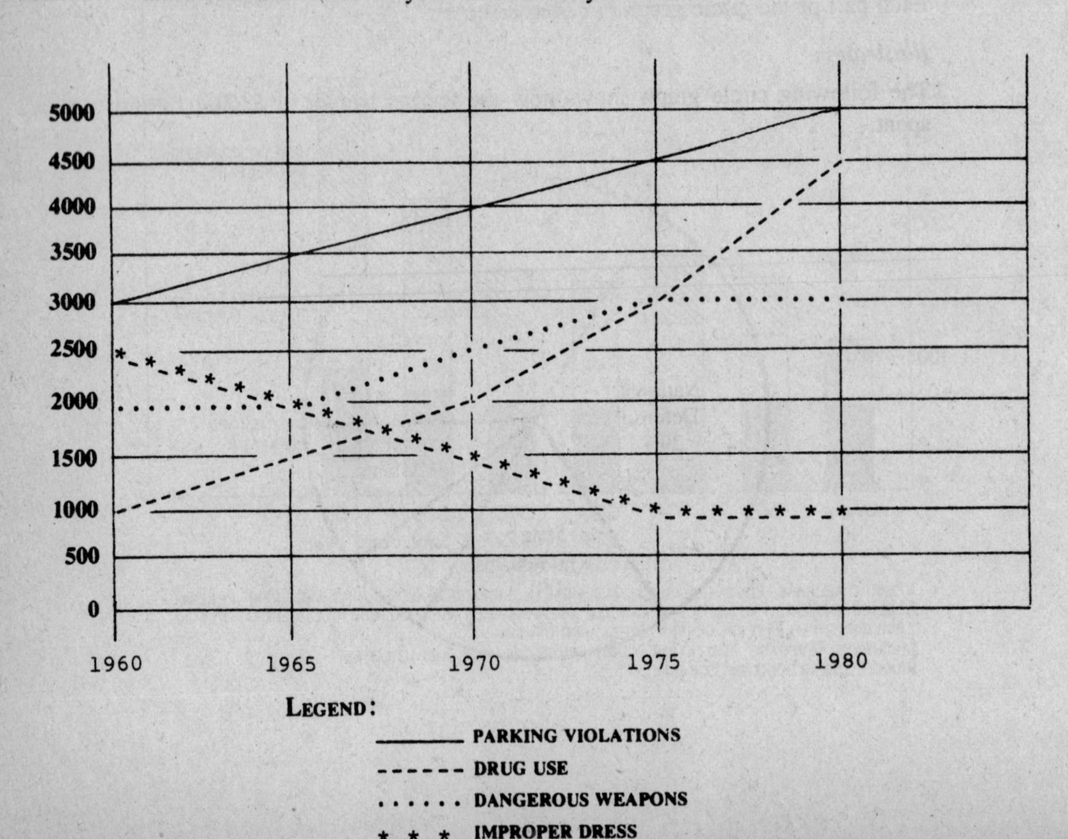

Question 4: Over the 20-year period, which offense shows an average rate of increase of more than 150 citations per year?

Answer: Drug use citations increased from 1000 in 1960 to 4500 in 1980. The average increase over the 20-year period is $\frac{3500}{20} = 175$.

Question 5: Over the 20-year period, which offense shows a constant rate of increase or decrease?

Answer: A straight line indicates a constant rate of increase or decrease. Of the four lines, the one representing parking violations is the only straight one.

Question 6: Which offense shows a total increase or decrease of 50% for the full 20-year period?

Answer: Dangerous weapons citations increased from 2000 in 1960 to 3000 in 1980, which is an increase of 50%.

Circle Graphs

6. **Circle graphs** are used to show the relationship of various parts of a quantity to each other and to the whole quantity.

7. Percents are often used in circle graphs. The 360 degrees of the circle represents 100%.

8. Each part of the circle graph is called a **sector**.

Illustration:

The following circle graph shows how the federal budget of $300.4 billion was spent.

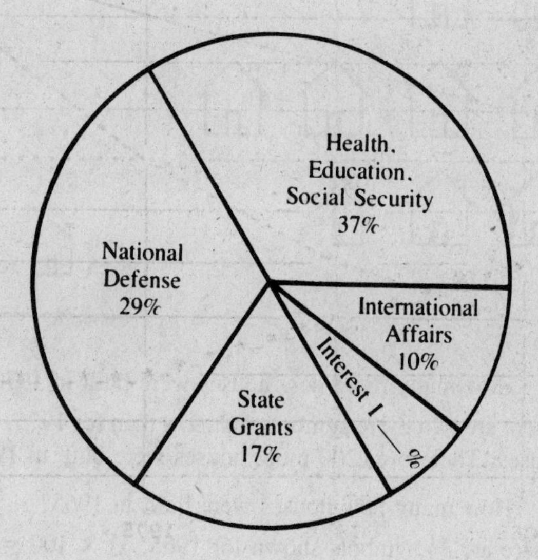

Question 7: What is the value of I?

Answer: There must be a total of 100% in a circle graph. The sum of the other sectors is:

$$17\% + 29\% + 37\% + 10\% = 93\%$$

Therefore, $I = 100\% - 93\% = 7\%$.

Question 8: How much money was actually spent on national defense?

Answer: $29\% \times \$300.4$ billion $= \$87.116$ billion
$= \$87{,}116{,}000{,}000$

Question 9: How much more money was spent on state grants than on interest?

Answer: $17\% - 7\% = 10\%$
$10\% \times \$300.4$ billion $= \$30.04$ billion
$= \$30{,}040{,}000{,}000$

Pictographs

9. **Pictographs** allow comparisons of quantities by using symbols. Each symbol represents a given number of a particular item.

Illustration:

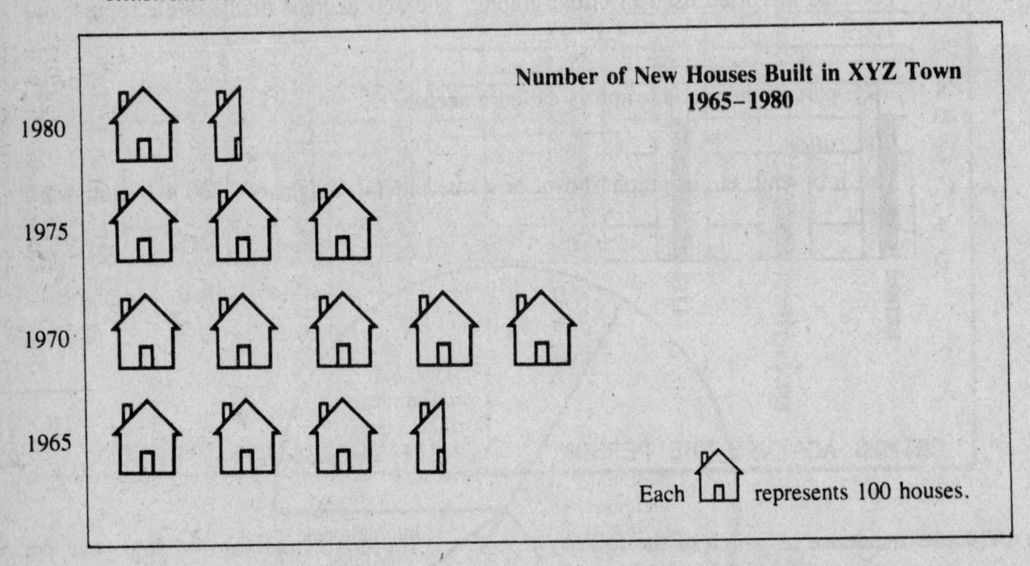

Question 10: How many more new houses were built in 1970 than in 1975?

Answer: There are two more symbols for 1970 than for 1975. Each symbol represents 100 houses. Therefore, 200 more houses were built in 1970.

Question 11: How many new houses were built in 1965?

Answer: There are $3\frac{1}{2}$ symbols shown for 1965; $3\frac{1}{2} \times 100 = 350$ houses.

Question 12: In which year were half as many houses built as in 1975?

Answer: In 1975, $3 \times 100 = 300$ houses were built. Half of 300, or 150, houses were built in 1980.

Practice Problems Involving Graphs

Questions 1–4 refer to the following graph:

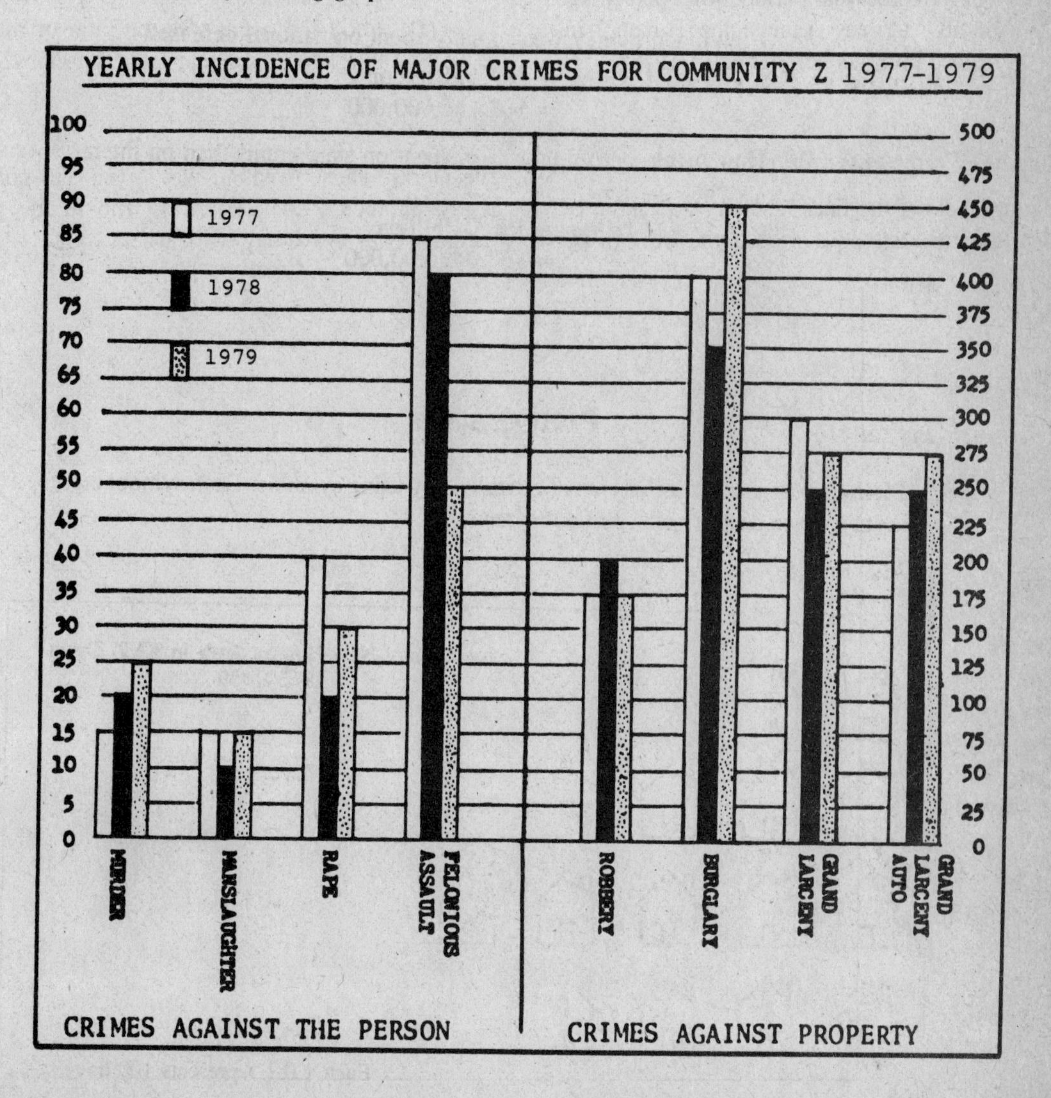

1. In 1979, the incidence of which of the following crimes was greater than in the previous two years?
   (A) grand larceny (C) rape
   (B) murder (D) robbery

2. If the incidence of burglary in 1980 had increased over 1979 by the same number as it had increased in 1979 over 1978, then the average for this crime for the four-year period from 1977 through 1980 would be most nearly
   (A) 100 (C) 425
   (B) 400 (D) 440

3. The above graph indicates that the *percentage* increase in grand larceny auto from 1978 to 1979 was:
   (A) 5% (C) 15%
   (B) 10% (D) 20%

4. Which of the following cannot be determined because there is not enough information in the above graph to do so?

(A) For the three-year period, what percentage of all "Crimes Against the Person" involved murders committed in 1978?
(B) For the three-year period, what percentage of all "Major Crimes" was committed in the first six months of 1978?
(C) Which major crimes followed a pattern of continuing yearly increases for the three-year period?
(D) For 1979, what was the ratio of robbery, burglary, and grand larceny crimes?

Questions 5–7 refer to the following graph:

In the graph below, the lines labeled "A" and "B" represent the cumulative progress in the work of two file clerks, each of whom was given 500 consecutively numbered applications to file in the proper cabinets over a five-day work week.

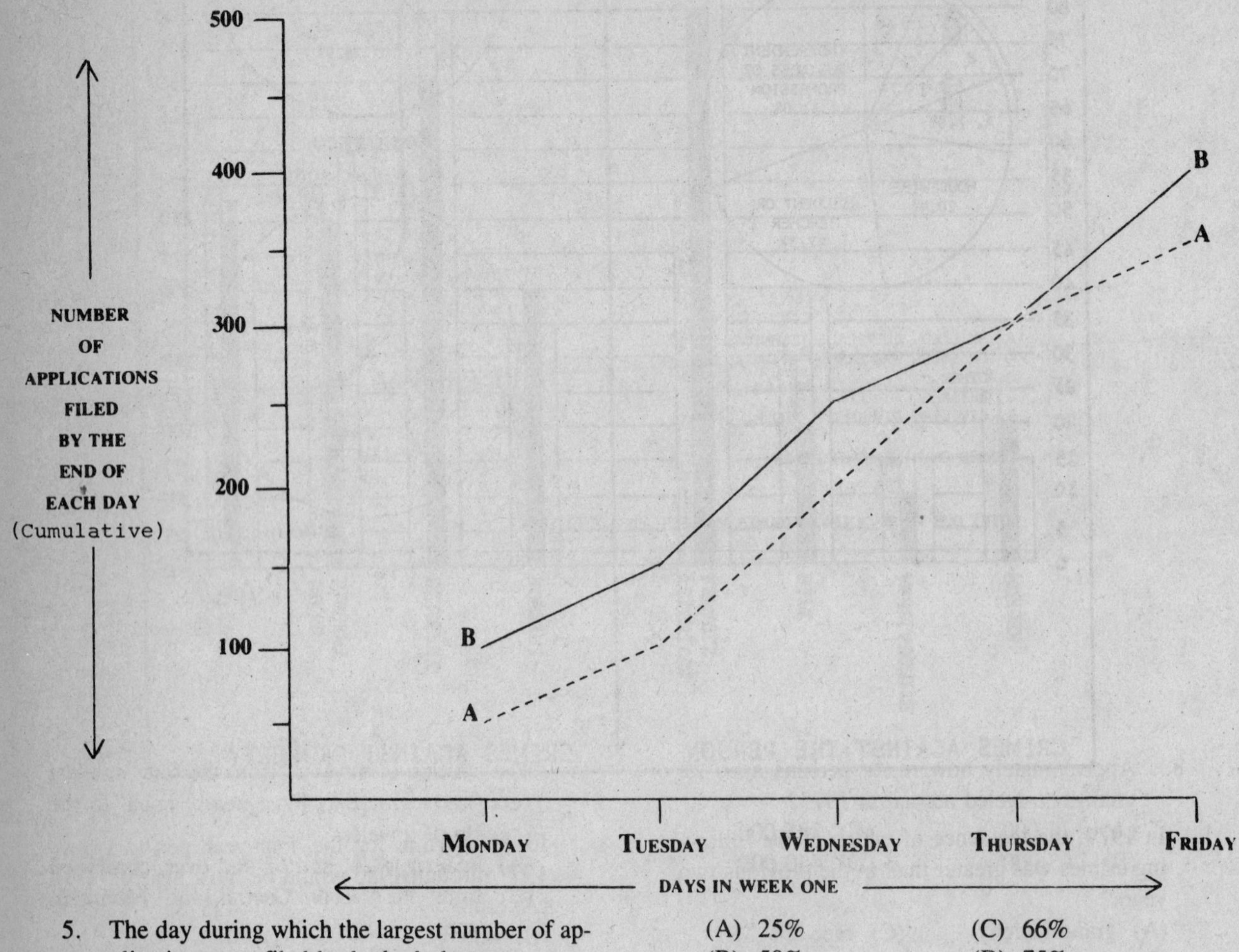

5. The day during which the largest number of applications was filed by both clerks was
(A) Monday
(B) Tuesday
(C) Wednesday
(D) Friday

6. At the end of the second day, the percentage of applications still to be filed was
(A) 25%
(B) 50%
(C) 66%
(D) 75%

7. Assuming that the production pattern is the same the following week as the week shown in the chart, the day on which Clerk B will finish this assignment will be
(A) Monday
(B) Tuesday
(C) Wednesday
(D) Friday

Questions 8–11 refer to the following graph:

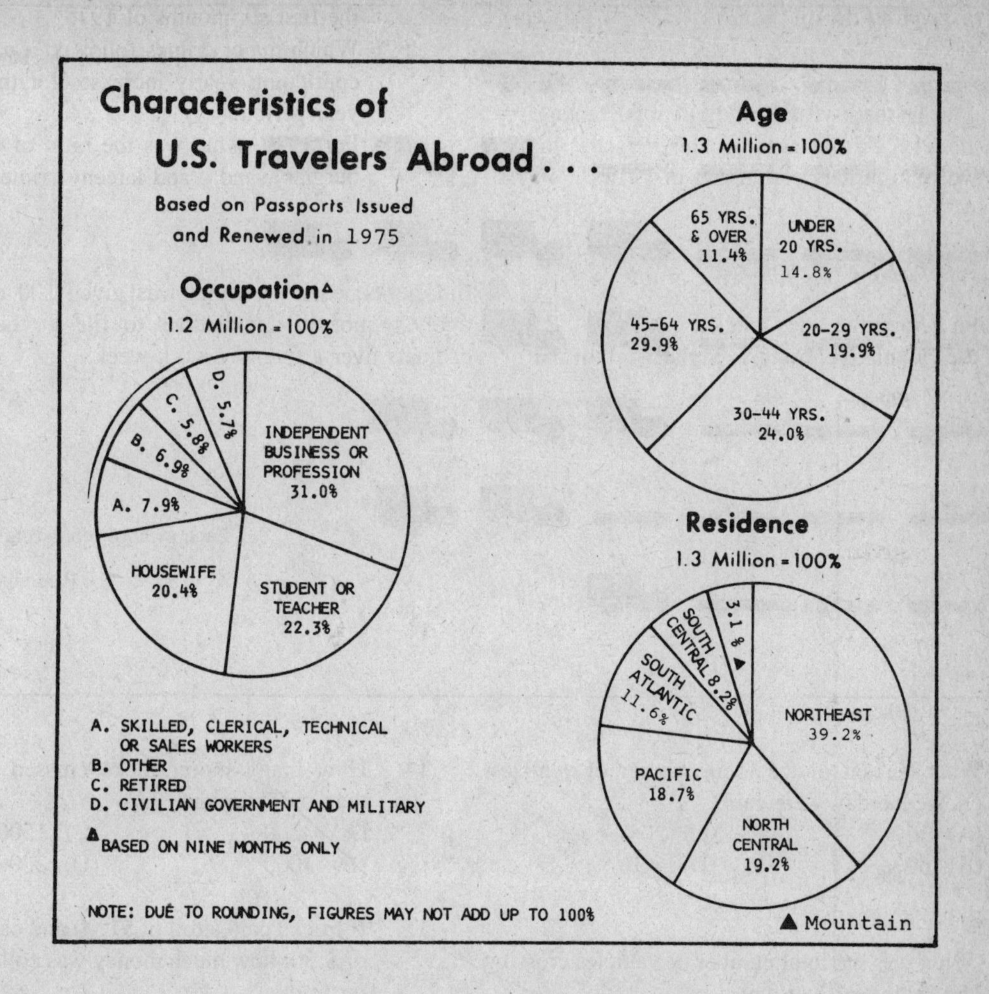

8. Approximately how many persons aged 29 or younger traveled abroad in 1975?
   (A) 175,000 (C) 385,000
   (B) 245,000 (D) 450,000

9. Of the people who did *not* live in the Northeast, what percent came from the North Central states?
   (A) 19.2% (C) 26.5%
   (B) 19.9% (D) 31.6%

10. The fraction of travelers from the four smallest occupation groups is most nearly equal to the fraction of travelers
    (A) under age 20, and 65 and over, combined
    (B) from the North Central and Mountain states
    (C) between 45 and 64 years of age
    (D) from the Housewife and Other categories

11. If the South Central, Mountain, and Pacific sections were considered as a single classification, how many degrees would its sector include?
    (A) 30° (C) 108°
    (B) 67° (D) 120°

Questions 12–15 refer to the following graph:

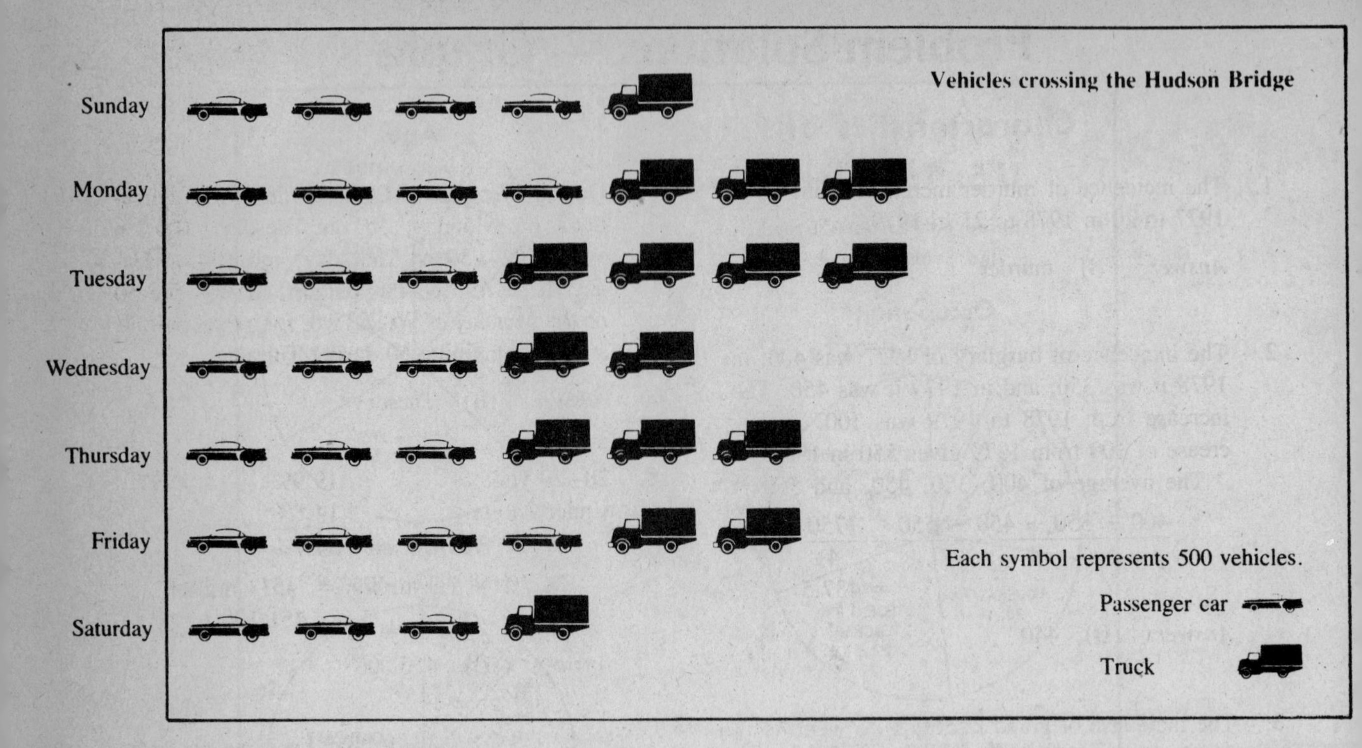

12. What percent of the total number of vehicles on Wednesday were cars?
(A) 30% (B) 60% (C) 20% (D) 50%

13. What was the total number of vehicles crossing the bridge on Tuesday?
(A) 7 (B) 700 (C) 1100 (D) 3500

14. How many more trucks crossed on Monday than on Saturday?
(A) 200 (B) 1000 (C) 1500 (D) 2000

15. If trucks paid a toll of $1.00 and cars paid a toll of $.50, how much money was collected in tolls on Friday?
(A) $400 (B) $600 (C) $2000 (D) $2500

Graphs — Correct Answers

1. (B)
2. (D)
3. (B)
4. (B)
5. (C)
6. (D)
7. (B)
8. (D)
9. (D)
10. (A)
11. (C)
12. (B)
13. (D)
14. (B)
15. (C)

Problem Solutions — Graphs

1. The incidence of murder increased from 15 in 1977 to 20 in 1978 to 25 in 1979.

   Answer: **(B)** murder

2. The incidence of burglary in 1977 was 400; in 1978 it was 350; and in 1979 it was 450. The increase from 1978 to 1979 was 100. An increase of 100 from 1979 gives 550 in 1980.
   The average of 400, 350, 450, and 550 is
   $$\frac{400 + 350 + 450 + 550}{4} = \frac{1750}{4} = 437.5$$

   Answer: **(D)** 440

3. The incidence of grand larceny auto went from 250 in 1978 to 275 in 1979, an increase of 25.
   The percent increase is
   $$\tfrac{25}{250} = .10 = 10\%$$

   Answer: **(B)** 10%

4. This graph gives information by year, not month. It is impossible to determine from the graph the percentage of crimes committed during the first six months of any year.

   Answer: **(B)**

5. For both A and B, the greatest increase in the cumulative totals occurred from the end of Tuesday until the end of Wednesday. Therefore, the largest number of applications was filed on Wednesday.

   Answer: **(C)** Wednesday

6. By the end of Tuesday, A had filed 100 applications and B had filed 150, for a total of 250. This left 750 of the original 1000 applications.
   $$\tfrac{750}{1000} = .75 = 75\%$$

   Answer: **(D)** 75%

7. During Week One, Clerk B files 150 applications on Monday, 50 on Tuesday, 100 on Wednesday, 50 on Thursday, and 100 on Friday. If he follows this pattern, he will file 50 on the Monday of Week Two, for a total of 450, and the remaining 50 during Tuesday.

   Answer: **(B)** Tuesday

8. 20–29 yrs.: 19.9%
   Under 20 yrs.: +14.8%
   Total: 34.7%

   $$34.7\% \times 1.3 \text{ million} = .4511 \text{ million} = 451{,}100$$

   Answer: **(D)** 450,000

9. 100% − 39.2% = 60.8% did not live in Northeast.
   19.2% lived in North Central
   $$\frac{19.2}{60.8} = .316 \text{ approximately}$$

   Answer: **(D)** 31.6%

10. Four smallest groups of occupation:
    $$7.9 + 6.9 + 5.8 + 5.7 = 26.3$$
    Age groups under 20 and over 65:
    $$14.8 + 11.4 = 26.2$$

    Answer: **(A)**

11. South Central: 8.2%
    Mountain: 3.1%
    Pacific: 18.7%
    Total: 30.0%

    $$30\% \times 360^\circ = 108^\circ$$

    Answer: **(C)** 108°

12. There are 5 vehicle symbols, of which 3 are cars. $\tfrac{3}{5} = 60\%$

    Answer: **(B)** 60%

13. On Tuesday, there were 3 × 500 = 1500 cars and 4 × 500 = 2000 trucks. The total number of vehicles was 3500.

    Answer: **(D)** 3500

14. The graph shows 2 more truck symbols on Monday than on Saturday. Each symbols represents 500 trucks, so there were 2 × 500 = 1000 more trucks on Monday.

    Answer: **(B)** 1000

15. On Friday there were

    4 × 500 = 2000 cars
    2 × 500 = 1000 trucks

Car tolls:	2000 × $ .50 =	$1000
Truck tolls:	1000 × $1.00 =	+ $1000
Total tolls:		$2000

Answer: **(C)** $2000

PAYROLL

1. **Salaries** are computed over various time periods: hourly, daily, weekly, biweekly (every 2 weeks), semimonthly (twice each month), monthly, and yearly.

2. **Overtime** is usually computed as "time and a half"; that is, each hour in excess of the number of hours in the standard workday or workweek is paid at $1\frac{1}{2}$ time the regular hourly rate. Some companies pay "double time," twice the regular hourly rate, for work on Sundays and holidays.

   Illustration: An employee is paid weekly, based on a 40-hour workweek, with time and a half for overtime. If the employee's regular hourly rate is $4.50, how much will he earn for working 47 hours in one week?

   SOLUTION: Overtime hours = 47 − 40 = 7 hours
   Overtime pay = $1\frac{1}{2} \times \$4.50 = \$6.75$ per hour

   Overtime pay for 7 hours:
   $7 \times \$6.75 = \$47.25$

   Regular pay for 40 hours:
   $40 \times \$4.50 = \$180.00$
   Total pay = $47.25 + $180 = $227.25

   Answer: $227.25

3. a. In occupations such as retail sales, real estate, and insurance, earnings may be based on **commission**, which is a percent of the sales or a percent of the value of the transactions that are completed.

   b. Earnings may be from straight commission only, from salary plus commission, or from a commission that is graduated according to transaction volume.

   Illustration: A salesman earns a salary of $200 weekly, plus a commission based on sales volume for the week. The commission is 7% for the first $1500 of sales and 10% for all sales in excess of $1500. How much did he earn in a week in which his sales totaled $3200?

   SOLUTION:

$3200 − $1500 =	$1700	excess sales
.07 × $1500 =	$105	commission on first $1500
.10 × $1700 =	$170	commission on excess sales
	+ $200	weekly salary
	$475	total earnings

   Answer: $475

4. **Gross pay** refers to the amount of money earned whether from salary, commission, or both, before any deductions are made.

5. There are several deductions that are usually made from gross pay:

   a. **Withholding tax** is the amount of money withheld for income tax. It is based on wages, marital status, and number of exemptions (also called allowances) claimed by the employee. The withholding tax is found by referring to tables supplied by the federal, state or city governments.

Example:

Married Persons — Weekly Payroll Period

Wages		Number of withholding allowances claimed				
At least	But less than	0	1	2	3	4
		Amount of income tax to be withheld				
400	410	73.00	67.60	62.30	57.70	53.10
410	420	75.80	70.40	65.00	60.10	55.50
420	430	78.60	73.20	67.80	62.50	57.90
430	440	81.40	76.00	70.60	65.20	60.30
440	450	84.20	78.80	73.40	68.00	62.70
450	460	87.00	81.60	76.20	70.80	65.40
460	470	90.20	84.40	79.00	73.60	68.20
470	480	93.40	87.30	81.80	76.40	71.00
480	490	96.60	90.50	84.60	79.20	73.80
490	500	99.80	93.70	87.50	82.00	76.60

Based on the above table, an employee who is married, claims three exemptions, and is paid a weekly wage of $434.50 will have $65.20 withheld for income tax. If the same employee earned $440 weekly it would be necessary to look on the next line for "at least $440 but less than $450" to find that $68.00 would be withheld.

   b. The FICA (Federal Insurance Contribution Act) tax is also called the Social Security tax. In 1982, the FICA tax was 6.7% of the first $32,400 of annual wages; the wages in excess of $32,400 were not subject to the tax.

   The FICA may be found by multiplying the wages up to and including $32,400 by .067, or by using tables such as the one below.

Example:

Social Security Employee Tax Table

6.7 percent employee tax deductions

Wages			Wages		
At least	But less than	Tax	At least	But less than	Tax
$78.14	$78.28	$5.24	$84.26	$84.40	$5.65
78.29	78.43	5.25	84.41	84.55	5.66
78.44	78.58	5.26	84.56	84.70	5.67
78.59	78.73	5.27	84.71	84.85	5.68
78.74	78.88	5.28	84.86	84.99	5.69
78.89	79.03	5.29	85.00	85.14	5.70
79.04	79.18	5.30	85.15	85.29	5.71
79.19	79.33	5.31	85.30	85.44	5.72
79.34	79.48	5.32	85.45	85.59	5.73
79.49	79.63	5.33	85.60	85.74	5.74
79.64	79.78	5.34	85.75	85.89	5.75
79.79	79.93	5.35			

According to the table above, the Social Security tax, or FICA tax, on wages of $84.80 is $5.20. The FICA tax on $84.92 is $5.21.

Illustration: Based on 1980 tax figures, what is the total FICA tax on an annual salary of $30,000?

SOLUTION: A maximum of $25,900 is taxable under FICA.

$$.0613 \times \$25{,}900 = \$1587.67$$

Answer: $1587.67

c. Other deductions that may be made from gross pay are deductions for pension plans, loan payments, payroll savings plans, and union dues.

6. The **net pay**, or **take-home pay**, is equal to gross pay less the total deductions.

Illustration: Mr. Jay earns $550 salary per week, with the following deductions: federal withholding tax, $106.70; FICA tax, $33.72; state tax, $22.83; pension payment, $6.42; union dues, $5.84. How much take-home pay does he receive?

SOLUTION: Deductions: $106.70
33.72
22.83
6.42
5.84
$175.51

Gross pay = $550.00
Deductions = − 175.51
Net pay = $374.49

Answer: His take-home pay is $374.49.

Practice Problems Involving Payroll

1. Jane Rose's semimonthly salary is $750. Her yearly salary is
(A) $9000
(B) 12,500
(C) $18,000
(D) $19,500

2. John Doe earns $300 for a 40-hour week. If he receives time and a half for overtime, what is his hourly overtime wage?
(A) $7.50
(B) $9.25
(C) $10.50
(D) $11.25

3. Which salary is greater?
(A) $350 weekly
(B) $1378 monthly
(C) $17,000 annually
(D) $646 biweekly

4. A factory worker is paid on the basis of an 8-hour day, with an hourly rate of $3.50 and time and a half for overtime. Find his gross pay for a week in which he worked the following hours: Monday, 8; Tuesday, 9; Wednesday, $9\frac{1}{2}$; Thursday, $8\frac{1}{2}$; Friday, 9.
(A) $140
(B) $154
(C) $161
(D) $231

Questions 5 and 6 refer to the following table:

Single Persons — Weekly Payroll Period

Wages		Number of withholding allowances claimed				
At least	But less than	0	1	2	3	4
		Amount of income tax to be withheld				
370	380	83.60	77.10	70.50	64.50	58.80
380	390	87.00	80.50	73.90	67.50	61.80
390	400	90.40	83.90	77.30	70.80	64.80
400	410	93.80	87.30	80.70	74.20	67.80
410	420	97.20	90.70	84.10	77.60	71.10
420	430	100.60	94.10	87.50	81.00	74.50
430	440	104.10	97.50	90.90	84.40	77.90
440	450	108.00	100.90	94.30	87.80	81.30
450	460	111.90	104.40	97.70	91.20	84.70
460	470	115.80	108.30	101.10	94.60	88.10

5. If an employee is single and has one exemption, the income tax withheld from his weekly salary of $389.90 is
(A) $87.00 (C) $80.50
(B) $90.40 (D) $83.90

6. If a single person with two exemptions has $90.90 withheld for income tax, his weekly salary could *not* be
(A) $430.00 (C) $437.80
(B) $435.25 (D) $440.00

7. Sam Richards earns $1200 monthly. The following deductions are made from his gross pay monthly: federal withholding tax, $188.40; FICA tax, $73.56; state tax, $36.78; city tax, $9.24; savings bond, $37.50; pension plan, $5.32; repayment of pension loan, $42.30. His monthly net pay is
(A) $806.90 (C) $808.90
(B) $807.90 (D) $809.90

8. A salesman is paid a straight commission that is 23% of his sales. What is his commission on $1260 of sales?
(A) $232.40 (C) $259.60
(B) $246.80 (D) $289.80

9. Ann Johnson earns a salary of $150 weekly plus a commission of 9% of sales in excess of $500 for the week. For a week in which her sales were $1496, her earnings were
(A) $223.64 (C) $253.64
(B) $239.64 (D) $284.64

10. A salesperson is paid a 6% commission on the first $2500 of sales for the week, and $7\frac{1}{2}$% on that portion of sales in excess of $2500. What is the commission earned in a week in which sales were $3280?
(A) $196.80 (C) $224.30
(B) $208.50 (D) $246.00

Payroll Problems — Correct Answers

1. (C)
2. (D)
3. (A)
4. (C)
5. (C)
6. (D)
7. (A)
8. (D)
9. (B)
10. (B)

Payroll Problems — Solutions

1. A semimonthly salary is paid twice a month. She receives $750 × 2 = $1500 each month, which is $1500 × 12 = $18,000 per year.

   Answer: (C) $18,000

2. The regular hourly rate is
   $300 ÷ 40 = $7.50
   The overtime rate is
   $7.50 × $1\frac{1}{2}$ = $7.50 × 1.5
   = $11.25

   Answer: (D) $11.25

3. Write each salary as its yearly equivalent:
   $350 weekly = $350 × 52 yearly
   = $18,200 yearly
   $1378 monthly = $1378 × 12 yearly
   = $16,536 yearly
   $17,000 annually = $17,000 yearly
   $646 biweekly = $646 ÷ 2 weekly
   = $323 weekly
   = $323 × 52 yearly
   = $16,796 yearly

   Answer: (A) $350 weekly

4. His overtime hours were:

Monday	0
Tuesday	1
Wednesday	$1\frac{1}{2}$
Thursday	$\frac{1}{2}$
Friday	1
Total	4 hours overtime

   Overtime rate per hour = $1\frac{1}{2}$ × $3.50
   = 1.5 × $3.50
   = $5.25
   Overtime pay = 4 × $5.25
   = $21

   Regular pay for 8 hours per day for 5 days or 40 hours.
   Regular pay = 40 × $3.50
   = $140
   Total wages = $140 + $21
   = $161

   Answer: (C) $161

5. The correct amount is found on the line for wages of at least $380 but less than $390, and in the column under "1" withholding allowance. The amount withheld is $80.50.

   Answer: (C) $80.50

6. In the column for 2 exemptions, or withholding allowances, $90.90 is found on the line for wages of at least $430, but less than $440. Choice (D) does not fall within that range.

   Answer: (D) $440

7. Deductions: $188.40
   73.56
   36.78
   9.24
   37.50
   5.32
   + 42.30
   Total $393.10

   Gross pay = $1200.00
   Total deductions = − 393.10
   $ 806.90

   Answer: (A) $806.90

8. 23% of $1260 = .23 × $1260
   = $289.80

   Answer: (D) $289.80

9. $1496 − 500 = $996 excess sales
   9% of $996 = .09 × $996
   = $89.64 commission

   $150.00 salary
   + 89.64 commission
   $239.64 total earnings

   Answer: (B) $239.64

10. $3280 − $2500 = $780 excess sales
    Commission on $2500:
    .06 × $2500 = $150.00
    Commission on $780:
    .075 × $780 = + 58.50
    Total = $208.50

    Answer: (B) $208.50

SEQUENCES

1. A **sequence** is a list of numbers based on a certain pattern. There are three main types of sequences:

   a. If each term in a sequence is being increased or diminished by the same number to form the next term, then it is an **arithmetic sequence**. The number being added or subtracted is called the **common difference**.

   Examples: 2, 4, 6, 8, 10 . . . is an arithmetic sequence in which the common difference is 2.

   14, 11, 8, 5, 2 . . . is an arithmetic sequence in which the common difference is 3.

   b. If each term of a sequence is being multiplied by the same number to form the next term, then it is a **geometric sequence**. The number multiplying each term is called the **common ratio**.

   Examples: 2, 6, 18, 54 . . . is a geometric sequence in which the common ratio is 3.

   64, 16, 4, 1 . . . is a geometric sequence in which the common ratio is $\frac{1}{4}$.

   c. If the sequence is neither arithmetic nor geometric, it is a **miscellaneous sequence**. Such a sequence may have each term a square or a cube, or the difference may be squares or cubes; or there may be a varied pattern in the sequence that must be determined.

2. A sequence may be ascending, that is, the numbers increase; or descending, that is, the numbers decrease.

3. To determine whether the sequence is arithmetic:

   a. If the sequence is ascending, subtract the first term from the second, and the second term from the third. If the difference is the same in both cases, the sequence is arithmetic.

   b. If the sequence is descending, subtract the second term from the first, and the third term from the second. If the difference is the same in both cases, the sequence is arithmetic.

4. To determine whether the sequence is geometric, divide the second term by the first, and the third term by the second. If the ratio is the same in both cases, the sequence is geometric.

5. To find a missing term in an arithmetic sequence that is ascending:

   a. Subtract any term from the one following it to find the common difference.

   b. Add the common difference to the term preceding the missing term.

   c. If the missing term is the first term, it may be found by subtracting the common difference from the second term.

Illustration: What number follows $16\frac{1}{3}$ in this sequence:

$$3,\ 6\tfrac{1}{3},\ 9\tfrac{2}{3},\ 13,\ 16\tfrac{1}{3}\ldots$$

SOLUTION: $6\frac{1}{3} - 3 = 3\frac{1}{3}$, $9\frac{2}{3} - 6\frac{1}{3} = 3\frac{1}{3}$
The sequence is arithmetic; the common difference is $3\frac{1}{3}$.

$$16\tfrac{1}{3} + 3\tfrac{1}{3} = 19\tfrac{2}{3}$$

Answer: The missing term, which is the term following $16\frac{1}{3}$, is $19\frac{2}{3}$.

6. To find a missing term in an arithmetic sequence that is descending:

   a. Subtract any term from the one preceding it to find the common difference.

   b. Subtract the common difference from the term preceding the missing term.

   c. If the missing term is the first term, it may be found by adding the common difference to the second term.

Illustration: Find the first term in the sequence:

$$\text{——},\ 16,\ 13\tfrac{1}{2},\ 11,\ 8\tfrac{1}{2},\ 6\ldots$$

SOLUTION: $16 - 13\frac{1}{2} = 2\frac{1}{2}$, $13\frac{1}{2} - 11 = 2\frac{1}{2}$
The sequence is arithmetic; the common difference is $2\frac{1}{2}$.

$$16 + 2\tfrac{1}{2} = 18\tfrac{1}{2}$$

Answer: The term preceding 16 is $18\frac{1}{2}$.

7. To find a missing term in a geometric sequence:

   a. Divide any term by the one preceding it to find the common ratio.

   b. Multiply the term preceding the missing term by the common ratio.

   c. If the missing term is the first term, it may be found by dividing the second term by the common ratio.

Illustration: Find the missing term in the sequence:

$$2,\ 6,\ 18,\ 54,\ \text{——}$$

SOLUTION: $6 \div 2 = 3$, $18 \div 6 = 3$
The sequence is geometric; the common ratio is 3.

$$54 \times 3 = 162$$

Answer: The missing term is 162.

Illustration: Find the missing term in the sequence:

——, 32, 16, 8, 4, 2

SOLUTION: $16 \div 32 = \frac{1}{2}$ (common ratio)

$32 \div \frac{1}{2} = 32 \times \frac{2}{1}$

$= 64$

Answer: The first term is 64.

8. If, after trial, a sequence is neither arithmetic nor geometric, it must be one of a miscellaneous type. Test to see whether it is a sequence of squares or cubes or whether the difference is the square or the cube of the same number; or the same number may be first squared, then cubed, etc.

Practice Problems Involving Sequences

Find the missing term in each of the following sequences:

1. ——, 7, 10, 13
2. 5, 10, 20, ——, 80
3. 49, 45, 41, ——, 33, 29
4. 1.002, 1.004, 1.006, ——
5. 1, 4, 9, 16, ——
6. 10, $7\frac{7}{8}$, $5\frac{3}{4}$, $3\frac{5}{8}$, ——
7. ——, 3, $4\frac{1}{2}$, $6\frac{3}{4}$
8. 55, 40, 28, 19, 13, ——
9. 9, 3, 1, $\frac{1}{3}$, $\frac{1}{9}$, ——
10. 1, 3, 7, 15, 31, ——

Sequence Problems — Correct Answers

1. 4
2. 40
3. 37
4. 1.008
5. 25
6. $1\frac{1}{2}$
7. 2
8. 10
9. $\frac{1}{27}$
10. 63

Problem Solutions — Sequences

1. This is an ascending arithmetic sequence in which the common difference is $10 - 7$, or 3. The first term is $7 - 3 = 4$.

2. This is a geometric sequence in which the common ratio is $10 \div 5$, or 2. The missing term is $20 \times 2 = 40$.

3. This is a descending arithmetic sequence in which the common difference is $49 - 45$, or 4. The missing term is $41 - 4 = 37$.

4. This is an ascending arithmetic sequence in which the common difference is $1.004 - 1.002$, or .002. The missing term is $1.006 + .002 = 1.008$.

5. This sequence is neither arithmetic nor geometric. However, if the numbers are rewritten as 1^2, 2^2, 3^2, and 4^2, it is clear that the next number must be 5^2, or 25.

6. This is a descending arithmetic sequence in which the common difference is $10 - 7\frac{7}{8} = 2\frac{1}{8}$. The missing term is $3\frac{5}{8} - 2\frac{1}{8} = 1\frac{4}{8}$, or $1\frac{1}{2}$.

7. This is a geometric sequence in which the common ratio is:

$$4\tfrac{1}{2} \div 3 = \tfrac{9}{2} \times \tfrac{1}{3} = \tfrac{3}{2}$$

The first term is $3 \div \frac{3}{2} = 3 \times \frac{2}{3} = 2$

Therefore, the missing term is 2.

8. There is no common difference and no common ratio in this sequence. However, note the differences between terms:

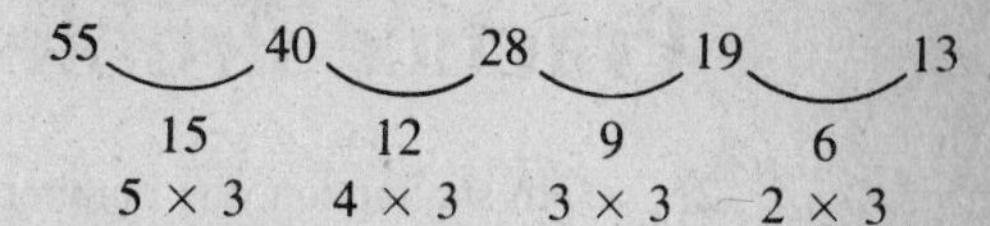

The differences are multiples of 3. Following the same pattern, the difference between 13 and the next term must be 1×3, or 3. The missing term is then $13 - 3 = 10$.

9. This is a geometric sequence in which the common ratio is $3 \div 9 = \frac{1}{3}$. The missing term is $\frac{1}{9} \times \frac{1}{3} = \frac{1}{27}$.

10. This sequence is neither arithmetic nor geometric. However, note the difference between terms:

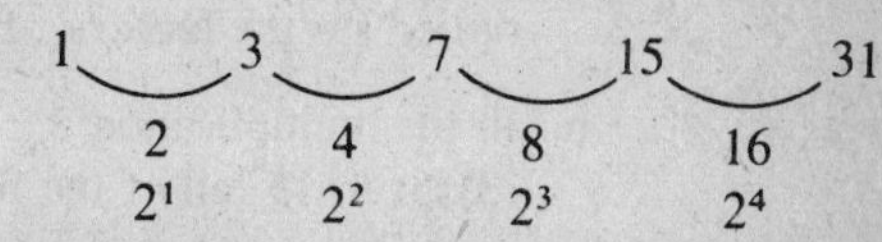

The difference between 31 and the next term must be 2^5, or 32. The missing term is $31 + 32 = 63$.

OPERATIONS WITH ALGEBRAIC EXPRESSIONS

Vocabulary

1. a. In addition, the numbers that are being added are called the **addends**. The solution to an addition problem is the **sum** or **total**.

   b. There are several ways to express an addition problem such as $10 + 2$:

the sum of 10 and 2	2 more than 10
the total of 10 and 2	2 greater than 10
2 added to 10	10 increased by 2

2. a. In subtraction, the number from which something is subtracted is the **minuend**, the number being subtracted is the **subtrahend** and the answer is the **difference**.

   Example: In $25 - 22 = 3$, the minuend is 25, the subtrahend is 22 and the difference is 3.

   b. A subtraction problem such as $25 - 22$ may be expressed as:

25 minus 22	from 25 take 22
25 less 22	25 decreased by 22
the difference of 25 and 22	22 less than 25
subtract 22 from 25	

3. a. In multiplication, the answer is called the **product** and the numbers being multiplied are the **factors** of the product.

   b. In the multiplication $3 \cdot 5 = 15$ [which may also be written as $3(5) = 15$ or $(3)(5) = 15$] all of the following expressions apply:

15 is the product of 3 and 5	15 is a multiple of 3
3 is a factor of 15	15 is a multiple of 5
5 is a factor of 15	

4. a. In division, the number being divided is the **dividend**, the number the dividend is divided by is the **divisor** and the answer is the **quotient**. Any number left over in the division is the **remainder**.

   Example: In $12 \div 2 = 6$, the dividend is 12, the divisor is 2 and the quotient is 6.

   Example: In $3\overline{)22}$ with quotient 7, 21, remainder 1:

   22 is the dividend
   3 is the divisor
   1 is the remainder

b. The division problem $12 \div 2$ may be expressed as:

12 divided by 2 — 2 divided into 12
the quotient of 12 and 2

Because $12 \div 2 = 6$ with no remainder, 2 is called a **divisor** of 12, and 12 is said to be **divisible** by 2.

Properties

5. Addition is a **commutative** operation; this means that two numbers may be added in either order without changing their sum:

$$2 + 3 = 3 + 2$$
$$a + b = b + a$$

Multiplication is also commutative:

$$4 \cdot 5 = 5 \cdot 4$$
$$ab = ba$$

6. Subtraction and division problems are *not* commutative; changing the order within a subtraction or division problem may affect the answer:

$$10 - 6 \neq 6 - 10$$
$$8 \div 4 \neq 4 \div 8$$

7. Addition and multiplication are **associative**; that is, if a problem involves only addition or only multiplication, the parentheses may be changed without affecting the answer. Parentheses are grouping symbols that indicate work to be done first.

$$(5 + 6) + 7 = 5 + (6 + 7)$$
$$(2 \cdot 3) \cdot 4 = 2 \cdot (3 \cdot 4)$$
$$(a + b) + c = a + (b + c)$$
$$(ab)c = a(bc)$$

8. Subtraction and division are *not* associative. Work within parentheses *must* be performed first.

$$(8 - 5) - 2 \neq 8 - (5 - 2)$$
$$(80 \div 4) \div 2 \neq 80 \div (4 \div 2)$$

9. a. Multiplication is **distributive** over addition. If a sum is to be multiplied by a number, instead of adding first and then multiplying, each addend may be multiplied by the number and the products added.

$$5(6 + 3) = 5 \cdot 6 + 5 \cdot 3$$
$$a(b + c) = ab + ac$$

b. Multiplication is also distributive over subtraction.

$$8(10 - 6) = 8 \cdot 10 - 8 \cdot 6$$
$$a(b - c) = ab - ac$$

c. The distributive property may be used in both directions.

$$5a + 3a = (5 + 3)a = 8a$$
$$847 \cdot 94 + 847 \cdot 6 = 847(94 + 6) = 847(100) = 84{,}700$$

Signed Numbers

10. a. A **signed number** is a number with a positive (+) or negative (−) sign in front of it. Signed numbers may be represented on a number line as follows:

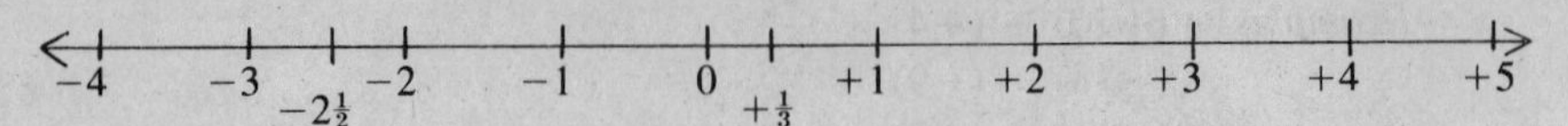

b. If a number (except zero) is written without a sign, it is assumed to be **positive**.

c. Zero is considered a signed number even though it is neither positive nor negative.

d. The magnitude, or **absolute value**, of a signed number is the number without its sign. The symbol used for absolute value is $|\ |$.

Examples: The absolute value of −3 is 3.

$|-3| = 3$

The absolute value of +6 is 6.

$|+6| = 6$

11. **Addition:**

a. To add two signed numbers that have the same sign, add their absolute values and give the answer the common sign.

Examples: $(+3) + (+4) = +7$
$(-6) + (-2) = -8$

b. To add two signed numbers that have different signs, subtract their absolute values. Give the answer the sign of the number with the *larger* absolute value.

Examples: $(-4) + (+1) = -3$
$(+5) + (-9) = -4$
$(-6) + (+7) = +1$

12. **Subtraction:**

To subtract two signed numbers, change the sign of the subtrahend. Then use the rules for addition of signed numbers.

Examples: $(-3) - (-5) = (-3) + (+5) = +2$
$(+10) - (-6) = (+10) + (+6) = +16$
$(+8) - (+9) = (+8) + (-9) = -1$
$(-7) - (+3) = (-7) + (-3) = -10$

13. **Multiplication:**

To multiply two signed numbers, multiply their absolute values. If the signed numbers have the same sign, the answer is positive. If the signed numbers have different signs, the answer is negative.

Examples: $(+3)(+4) = +12$
$(-5)(-2) = +10$
$(-6)(+3) = -18$
$(+8)(-1) = -8$

14. **Division:**

To divide two signed numbers, divide their absolute values. If the signed numbers have the same sign, the answer is positive. If the signed numbers have different signs, the answer is negative.

Examples:
$(+20) \div (+4) = +5$
$(-18) \div (-9) = +2$
$(-14) \div (+2) = -7$
$(+15) \div (-5) = -3$

15. To evaluate algebraic expressions and formulas:

a. Substitute the given values for the letters in the expression.

b. Perform the arithmetic in the following order:

First, perform the operations within parentheses (if any);

Second, compute all powers and roots;

Third, perform all multiplications and divisions in order from left to right;

Fourth, perform all additions and subtractions in order from left to right.

Illustration: If $P = 2(L + W)$, find P when $L = 10$ and $W = 5$

SOLUTION: Substitute 10 for L and 5 for W:

$P = 2(10 + 5)$ — First, add numbers in parentheses.
$= 2(15)$ — Then multiply 2 by 15.
$= 30$

Answer: 30

Illustration: Evaluate $5a^2 - 2b$ if $a = 3$ and $b = 10$

SOLUTION: Substitute 3 for a and 10 for b:

$5 \cdot 3^2 - 2 \cdot 10$ — First, find 3^2.
$5 \cdot 9 - 2 \cdot 10$ — Next, multiply $5 \cdot 9$ and $2 \cdot 10$.
$45 - 20$ — Then subtract 20 from 45.
25

Answer: 25

16. a. Algebraic expressions may contain numbers (constants) or letters (variables) or both.

b. In an algebraic expression, if several quantities are being added or subtracted, each of these quantities is called a **term**.

Example: In $4x^2 + 5y + 6$, the terms are: $4x^2$, $5y$, 6.

c. The number factor of each term is called the **coefficient**. The letter part is called the **literal factor**.

Example: In $3x^2$, 3 is the coefficient and x is the literal factor. Note that 2 is the exponent and is part of the literal factor.

d. Any variable appearing without a coefficient is assumed to have a coefficient of 1: $b = 1b$

e. Any variable appearing without an exponent is assumed to have an exponent of 1: $b = b^1$

17. a. If two or more terms have identical literal factors, they are called **like terms**.

Example: $3a$, $6a$ and a are like terms
$2x^4$ and $5x^2$ are *not* like terms

b. Terms may be added (or subtracted) only if they are like terms. Add (or subtract) the coefficients and repeat the literal factor. This is called **combining like terms**.

Examples:

$$3d + 2d = 5d$$
$$6xy + (-4)xy = 2xy$$
$$10z^3 + 5z^3 - 8z^3 = 7z^3$$

c. In most algebraic expressions it is easier to consider the operation that separates terms to be addition only, and the + or − sign immediately preceding each term to be the sign of the coefficient of that term.

Polynomials

18. An expression containing a single term is called a **monomial**. An expression containing more than one term is called a **polynomial**. Special polynomials are **binomials** (two terms) and **trinomials** (three terms).

19. To add (or subtract) two polynomials, add (or subtract) the coefficients of the like terms and repeat the literal factors. The unlike terms may *not* be combined.

Examples: Add

$$\begin{array}{r} 4x^2 - 3x + 2 \\ 2x^2 - 7x - 5 \\ \hline 6x^2 - 10x - 3 \end{array}$$

Subtract

$$\begin{array}{r} 7a - 2b + 4c \\ 9a + 6b - 2c \\ \hline -2a - 8b + 6c \end{array}$$

(Recall that in subtraction the sign of the subtrahend is changed and the rules of addition are used.)

20. To multiply two monomials, multiply their coefficients and add the exponents of like variables.

Examples:

$$2x^5 \cdot 3x^4 = 6x^9$$
$$y^4 \cdot y^{10} = y^{14}$$
$$9b^3 \cdot 2b = 18b^4 \qquad \text{(Note that } 2b = 2b^1\text{)}$$
$$(-4a^2b^3)(-3a^{11}b^8) = +12a^{13}b^{11}$$

21. To multiply a polynomial by a monomial, use the distributive property and multiply each term of the polynomial by the monomial.

Examples:

$$3(2x + 4y) = 6x + 12y$$
$$y^2(5y - 3y^5) = 5y^3 - 3y^7$$

22. To multiply a polynomial by a polynomial, multiply each term of the first polynomial by each term of the second polynomial, then add any like terms in the answer.

Examples:

$$(x + 3)(x + 4) = x^2 + 4x + 3x + 12$$
$$= x^2 + 7x + 12$$
$$(a - 1)(b + 5) = ab + 5a - 1b - 5$$
$$(y + 4)(y^2 + 2y - 3) = y^3 + 2y^2 - 3y + 4y^2 + 8y - 12$$
$$= y^3 + 6y^2 + 5y - 12$$

23. To divide two monomials, divide their coefficients and subtract the exponents of like variables.

Examples: $\frac{12a^5}{3a^2} = 4a^3$

$\frac{ac^7}{ac^5} = c^2$ (Note that $\frac{a}{a} = 1$)

$\frac{-6b^{10}c^7}{+2bc^2} = -3b^9c^5$

24. To divide a polynomial by a monomial, divide each term of the polynomial by the monomial.

Examples: $\frac{15a^2 - 12a}{3} = 5a^2 - 4a$

$(12x^3 - 8x^2 + 20x) \div 4x = 3x^2 - 2x + 5$

Simplifying Algebraic Expressions

25. Algebraic expressions containing parentheses can be simplified by using the following rules:

a. If a positive (+) sign is immediately before the parentheses, the parentheses may simply be omitted.

Example: $3x + (2y + z) = 3x + 2y + z$

b. If a negative (−) sign immediately precedes parentheses, the sign of each term within the parentheses must be changed. The parentheses may then be omitted.

Example: $4 - (2x - y + z) = 4 - 2x + y - z$

c. If a number or letter is indicated as a multiplier immediately before parentheses, the distributive property is used to multiply each term inside the parentheses by the multiplier.

Example: $a - 3(b + c) = a - 3b - 3c$

d. After removing parentheses, combine like terms.

Example: $5z + 2(3z - 4) = 5z + 6z - 8$
$= 11z - 8$

Factoring

26. To **factor** an expression means to find those quantities whose product is the original expression.

27. **Common factors:**

If all of the terms of a polynomial have a common factor, the distributive property may be used.

Examples: $ax + ay = a(x + y)$

$12d - 8f = 4(3d - 2f)$

$x^3 + 2x^2 - 4x = x(x^2 + 2x - 4)$

28. **Difference of two squares:**

A binomial which is the difference of two squares has as its factors two binomials, one the sum of the square roots, the other the difference of the square roots.

Examples: $x^2 - 9 = (x + 3)(x - 3)$
$25 - y^2 = (5 + y)(5 - y)$

29. **Trinomials:**

a. Quadratic trinomials are of the form: $ax^2 + bx + c$, where a, b and c are constants and $a \neq 0$. Some — but not all — quadratic trinomials can be factored into two binomials, each the sum of an x term and a numerical term.

b. When $a = 1$, the trinomial is written $x^2 + bx + c$. Each binomial factor will be the sum of x and a number. The product of the numbers is c; their sum is b.

Illustration: Factor $x^2 + 7x + 12$

SOLUTION: The product of the numerical parts of the factors must be 12. Pairs of numbers whose product is 12 are:

1 and 12
2 and 6
3 and 4

Of these pairs, the only one whose sum is 7 is 3 and 4. Therefore, the factors are $(x + 3)$ and $(x + 4)$.

Answer: $x^2 + 7x + 12 = (x + 3)(x + 4)$

Illustration: Factor $y^2 + 5y - 6$

SOLUTION: Pairs of numbers whose product is -6 are:

-1 and $+6$
$+1$ and -6
$+2$ and -3
-2 and $+3$

The pair whose sum is $+5$ is -1 and $+6$. Therefore, the factors are $(y - 1)$ and $(y + 6)$.

Answer: $y^2 + 5y - 6 = (y - 1)(y + 6)$

Illustration: Factor $z^2 - 11z + 10$

SOLUTION: The numbers whose product is positive are either both positive or both negative. In this case the sum of the numbers is negative, so consider only the negative pairs. The pairs of negative numbers whose product is $+10$ are:

-1 and -10
-2 and -5

The pair with -11 as its sum is -1 and -10. Therefore the factors are $(z - 1)$ and $(z - 10)$.

Answer: $z^2 - 11z + 10 = (z - 1)(z - 10)$

c. When $a \neq 1$ in the trinomial $ax^2 + bx + c$, the product of the x terms in the binomial factors must be ax^2, the product of the number terms must be c, and when the binomials are multiplied their product must be $ax^2 + bx + c$.

While there will be more than one possible pair of factors in which the product of the number terms is c, the correct pair is the only one whose product is the original trinomial.

Illustration: Factor $3x^2 + 10x + 8$

SOLUTION: The possible pairs of factors to be considered are:

$$(3x + 1)(x + 8)$$
$$(3x + 8)(x + 1)$$
$$(3x + 2)(x + 4)$$
$$(3x + 4)(x + 2)$$

In each case the product of the x terms is $3x^2$ and the product of the number terms is 8. Since the middle term is positive, any negative possibilities are ignored. Multiplying each pair of factors gives:

$$(3x + 1)(x + 8) = 3x^2 + 24x + 1x + 8$$
$$= 3x^2 + 25x + 8$$
$$(3x + 8)(x + 1) = 3x^2 + 3x + 8x + 8$$
$$= 3x^2 + 11x + 8$$
$$(3x + 2)(x + 4) = 3x^2 + 12x + 2x + 8$$
$$= 3x^2 + 14x + 8$$
$$(3x + 4)(x + 2) = 3x^2 + 6x + 4x + 8$$
$$= 3x^2 + 10x + 8$$

Therefore, $3x^2 + 10x + 8$ may be factored as $(3x + 4)(x + 2)$.

30. An expression may require more than one type of factoring before it is factored completely. To factor *completely*:

a. Use the distributive property to remove the highest common factor from each term.

b. If possible, factor the resulting polynomial as the difference of two squares or as a quadratic trinomial.

Examples:

$$3x^2 - 48 = 3(x^2 - 16)$$
$$= 3(x + 4)(x - 4)$$
$$2ay^2 + 12ay - 14a = 2a(y^2 + 6y - 7)$$
$$= 2a(y + 7)(y - 1)$$

Radicals

31. a. The symbol $\sqrt{x}$ means the positive square root of x. The $\sqrt{\ }$ is called the **radical sign** and the x is called the **radicand**.

b. Any positive number has two square roots, one positive and one negative. $\sqrt{x}$ indicates the positive root and $-\sqrt{x}$ indicates the negative root.

Example: The square roots of 100 are 10 and -10, since $(10)^2 = 100$ and $(-10)^2 = 100$.

$$\sqrt{100} = 10$$
$$-\sqrt{100} = -10$$

32. Many radicals may be simplified by using the principle $\sqrt{ab} = \sqrt{a} \cdot \sqrt{b}$

Examples: $\sqrt{100} = \sqrt{25}\sqrt{4} = 5 \cdot 2 = 10$
$\sqrt{18} = \sqrt{9}\sqrt{2} = 3\sqrt{2}$
$\sqrt{75} = \sqrt{25}\sqrt{3} = 5\sqrt{3}$

Note that the factors chosen must include at least one perfect square.

33. a. Radicals with the same radicands may be added or subtracted as like terms.

Examples: $3\sqrt{5} + 4\sqrt{5} = 7\sqrt{5}$
$10\sqrt{2} - 6\sqrt{2} = 4\sqrt{2}$

b. Radicals with different radicands may be combined only if they can be simplified to have like radicands.

Example: $\sqrt{50} + \sqrt{32} - 2\sqrt{2} + \sqrt{3} = \sqrt{25}\sqrt{2} + \sqrt{16}\sqrt{2} - 2\sqrt{2} + \sqrt{3}$
$= 5\sqrt{2} + 4\sqrt{2} - 2\sqrt{2} + \sqrt{3}$
$= 7\sqrt{2} + \sqrt{3}$

34. To multiply radicals, first multiply the coefficients. Then multiply the radicands.

Example: $2\sqrt{3} \cdot 4\sqrt{5} = 8\sqrt{15}$

35. To divide radicals, first divide the coefficients. Then divide the radicands.

Example: $\dfrac{14\sqrt{20}}{2\sqrt{2}} = 7\sqrt{10}$

Summary of Kinds of Numbers

36. The numbers that have been used in this section are called **real numbers** and may be grouped into special categories.

a. The **natural** numbers, or counting numbers, are:

1, 2, 3, 4, 5, 6, 7, 8, 9, 10, 11, 12, . . .

b. A natural number (other than 1) is a **prime** number if it can be exactly divided only by itself and 1. If a natural number has other divisors it is a **composite** number. The numbers 2, 3, 5, 7, and 11 are prime numbers, while 4, 6, 8, 9 and 12 are composites.

c. The **whole** numbers consist of 0 and the natural numbers:

0, 1, 2, 3, . . .

d. The **integers** consist of the natural numbers, the negatives of the natural numbers, and zero:

. . . −3, −2, −1, 0, 1, 2, 3, 4, . . .

Even integers are exactly divisible by 2:

. . . −6, −4, −2, 0, 2, 4, 6, 8, . . .

Odd integers are not divisible by 2:

. . . −5, −3, −1, 1, 3, 5, 7, 9, . . .

e. The **rational** numbers are numbers that can be expressed as the quotient of two integers (excluding division by 0). Rational numbers include integers, fractions, terminating decimals (such as 1.5 or .293) and repeating decimals (such as .333 . . . or .74867676767 . . .).

f. The **irrational** numbers cannot be expressed as the quotient of two integers, but can be written as non-terminating, non-repeating decimals. The numbers $\sqrt{2}$ and π are irrational.

Algebraic Fractions

37. To reduce an algebraic fraction, divide both the numerator and the denominator by the same *factor*.

Examples:

$$\frac{3x^2}{xy} = \frac{3\cancel{x}x}{\cancel{x}y} = \frac{3x}{y}$$

$$\frac{3a - 9}{a^2 - 9} = \frac{3\cancel{(a - 3)}}{\cancel{(a - 3)}(a + 3)} = \frac{3}{a + 3}$$

Note that $\frac{3}{a + 3} \neq \frac{\cancel{3}}{a + \cancel{3}}$ since 3 is a term, not factor, of a + 3.

38. To multiply two or more fractions, first cancel those factors which are common to any numerator and any denominator. Then multiply the numerators and multiply the denominators, using all remaining factors.

Example: $\frac{2x^2}{6(x + 3)} \cdot \frac{(x + 3)}{x^2y}$

$$\frac{\cancel{2x^2}}{\cancel{6}\cancel{(x + 3)}} \cdot \frac{\cancel{(x + 3)}}{\cancel{x^2}y}$$ Cancel the factors 2, x^2 and x + 3

$$\frac{1}{3} \cdot \frac{1}{y}$$ Multiply the remaining factors

$$\frac{1}{3y}$$ Answer

39. To divide two fractions, multiply the first fraction by the reciprocal of the second fraction.

Example: $\frac{x^2 - 3x + 2}{x^2} \div \frac{x^2 - 4}{x^3}$

$$\frac{x - 3x + 2}{x^2} \cdot \frac{x^3}{x^2 - 4}$$ Multiply by the reciprocal of the second fraction

$$\frac{\cancel{(x - 2)}(x - 1)}{\cancel{xx}} \cdot \frac{\cancel{xx}x}{\cancel{(x - 2)}(x + 2)}$$ Write the numerators and denominators as factors and cancel

$$\frac{1 \cdot (x - 1)}{1} \cdot \frac{x}{1 \cdot (x + 2)}$$ Multiply remaining factors in numerators and in denominators

$$\frac{x^2 - x}{x + 2}$$ Answer

40. a. To add (or subtract) fractions that have the same denominator, add (or subtract) their numerators. The denominator of the answer is the same as the denominator of the original fractions. Reduce the answer to lowest terms.

Examples: $\frac{2a}{3x} + \frac{5a}{3x} + \frac{2a}{3x} = \frac{\overset{3}{\cancel{9}}a}{\underset{1}{\cancel{3}}x} = \frac{3a}{x}$

$$\frac{2x}{y^2} - \frac{x-3}{y^2} = \frac{2x-(x-3)}{y^2}$$

$$= \frac{2x - x + 3}{y^2}$$

$$= \frac{x+3}{y^2}$$

b. To add (or subtract) two fractions that have different denominators, first change them to equivalent fractions having the same denominators, as follows:

1. Factor the denominator of each fraction, if possible.
2. Determine which factors in the second denominator are missing from the first denominator and multiply the numerator and denominator of the first fraction by the missing factors.
3. Determine which factors in the first denominator are missing from the second denominator, and multiply the numerator and denominator of the second fraction by the missing factors.
4. The fractions now have the same denominators and may be added or subtracted.

Examples: $\frac{2}{ab} + \frac{3}{bc}$

The denominator of the first fraction is missing a factor of c. Multiply numerator and denominator by c.

The denominator of the second fraction is missing a factor of a. Multiply numerator and denominator by a.

$$\frac{c \cdot 2}{c \cdot ab} + \frac{a \cdot 3}{a \cdot bc} = \frac{2c}{abc} + \frac{3a}{abc} = \frac{2c+3a}{abc}$$

$$\frac{6x}{x^2-25} - \frac{4}{x-5}$$

$= \frac{6x}{(x-5)(x+5)} - \frac{4}{x-5}$ Factor the denominators where possible

$= \frac{6x}{(x-5)(x+5)} - \frac{4}{(x-5)} \cdot \frac{(x+5)}{(x+5)}$ Only the second denominator is missing a factor. Multiply numerator and denominator by x + 5

$= \frac{6x - 4(x+5)}{(x-5)(x+5)}$ Combine numerators

$$= \frac{6x - 4x - 20}{(x-5)(x+5)}$$

$$= \frac{2x-20}{(x-5)(x+5)} \text{ or } \frac{2(x-10)}{(x-5)(x+5)}$$

41. **Rationalizing denominators:**

To simplify a fraction that has a radical denominator, multiply numerator and denominator by the radical, then reduce if possible.

Example: $\frac{ab}{\sqrt{a}} = \frac{ab}{\sqrt{a}} \cdot \frac{\sqrt{a}}{\sqrt{a}} = \frac{\overset{1}{\cancel{a}}b\sqrt{a}}{\underset{1}{\cancel{a}}} = b\sqrt{a}$

Practice Problems — Algebraic Expressions

1. The value of $2(-3) - |-4|$ is
   (A) -10 (C) 2
   (B) -2 (D) 10

2. The value of $3a^2 + 2a - 1$ when $a = -1$ is
   (A) -3 (C) 3
   (B) 0 (D) 6

3. If $2x^4$ is multiplied by $7x^3$ the product is
   (A) $9x^7$ (C) $14x^7$
   (B) $9x^{12}$ (D) $14x^{12}$

4. The expression $3(x - 4) - (3x - 5) + 2(x + 6)$ is equivalent to
   (A) $2x - 15$ (C) $2x + 5$
   (B) $2x + 23$ (D) $-2x - 15$

5. The product of $(x + 5)$ and $(x + 5)$ is
   (A) $2x + 10$ (C) $x^2 + 10x + 25$
   (B) $x^2 + 25$ (D) $x^2 + 10$

6. The quotient of $(4x^3 - 2x^2) \div (x^2)$ is
   (A) $4x^3 - 1$ (C) $4x^5 - 2x^4$
   (B) $4x - 2x^2$ (D) $4x - 2$

7. The expression $(+3x^4)^2$ is equal to
   (A) $6x^8$ (C) $9x^8$
   (B) $6x^6$ (D) $9x^6$

8. If $3x - 1$ is multiplied by $2x$, the product is
   (A) $4x$ (C) $6x^2 - 1$
   (B) $5x^2$ (D) $6x^2 - 2x$

9. One factor of the trinomial $x^2 - 3x - 18$ is
   (A) $x - 9$ (C) $x - 3$
   (B) $x - 6$ (D) $x + 9$

10. The sum of $\sqrt{18}$ and $\sqrt{72}$ is
    (A) $18\sqrt{2}$ (C) $3\sqrt{10}$
    (B) $9\sqrt{2}$ (D) 40

11. Reduce $\frac{2x + 10}{x^2 + 7x + 10}$ to lowest terms.
    (A) $\frac{1}{4x^2}$ (C) $\frac{2}{x + 2}$
    (B) $\frac{1}{x}$ (D) $\frac{2x}{x^2 + 7x}$

12. $\frac{4x + 3}{10x} + \frac{1}{2x} - \frac{2}{5} =$
    (A) 1 (C) $\frac{4}{5x}$
    (B) $\frac{16}{5}$ (D) $\frac{2x + 3}{7x}$

13. $\frac{x}{x + 7} \cdot (2x + 14) =$
    (A) $2x$ (C) $\frac{x}{2}$
    (B) $2x + 2$ (D) $\frac{x}{2(x + 7)}$

14. $\frac{3x}{5y} \div \frac{x}{y} =$
    (A) $\frac{3x^2}{5y^2}$ (C) $\frac{3y}{5x}$
    (B) $\frac{5y^2}{3x^2}$ (D) $\frac{3}{5}$

15. Simplify $\frac{20\sqrt{3}}{\sqrt{5}}$
    (A) $4\sqrt{3}$ (C) $100\sqrt{3}$
    (B) $4\sqrt{15}$ (D) $10\sqrt{3}$

Practice Problems — Correct Answers

1. **(A)**
2. **(B)**
3. **(C)**
4. **(C)**
5. **(C)**
6. **(D)**
7. **(C)**
8. **(D)**
9. **(B)**
10. **(B)**
11. **(C)**
12. **(C)**
13. **(A)**
14. **(D)**
15. **(B)**

Problem Solutions — Algebraic Expressions

1. $$2(-3) - |-4| = -6 - 4$$
$$= -10$$

Recall that $|-4|$ means the *absolute value* of -4, which is 4.

Answer: **(A)** -10

2. If $a = -1$,

$$3a^2 + 2a - 1 = 3(-1)^2 + 2(-1) - 1$$
$$= 3(+1) + 2(-1) - 1$$
$$= 3 - 2 - 1$$
$$= 0$$

Answer: **(B)** 0

3. $(2x^4)(7x^3) = 14x^7$ To multiply monomials, multiply coefficients and add exponents of like variables.

Answer: **(C)** $14x^7$

4. $$3(x - 4) - (3x - 5) + 2(x + 6)$$
$$= 3x - 12 - 3x + 5 + 2x + 12$$
$$= 2x + 5$$

Answer: **(C)** $2x + 5$

5. $$(x + 5)(x + 5) = x^2 + 5x + 5x + 25$$
$$= x^2 + 10x + 25$$

Answer: **(C)** $x^2 + 10x + 25$

6. $$(4x^3 - 2x^2) \div x^2 = 4x^3 \div x^2 - 2x^2 \div x^2$$
$$= 4x - 2$$

Answer: **(D)** $4x - 2$

7. $$(+3x^4)^2 = (+3x^4)(+3x^4)$$
$$= 9x^8$$

Answer: **(C)** $9x^8$

8. $$2x(3x - 1) = 2x \cdot 3x - 2x \cdot 1$$
$$= 6x^2 - 2x$$

Answer: **(D)** $6x^2 - 2x$

9. Factor $x^2 - 3x - 18$ by finding two numbers whose product is -18 and whose sum is -3. Pairs of numbers whose product is -18 are:

-1 and $+18$
$+1$ and -18
-9 and $+2$
$+9$ and -2
-6 and $+3$
$+6$ and -3

Of these pairs, the one whose sum is -3 is -6 and $+3$. Therefore, the factors of $x^2 - 3x - 18$ are $(x - 6)$ and $(x + 3)$.

Answer: **(B)** $x - 6$

10. $$\sqrt{18} + \sqrt{72} = \sqrt{9}\sqrt{2} + \sqrt{36}\sqrt{2}$$
$$= 3\sqrt{2} + 6\sqrt{2}$$
$$= 9\sqrt{2}$$

Answer: **(B)** $9\sqrt{2}$

11. $$\frac{2x + 10}{x^2 + 7x + 10} = \frac{2\cancel{(x + 5)}^{1}}{(x + 2)\cancel{(x + 5)}_{1}} = \frac{2}{x + 2}$$

Answer: **(C)** $\frac{2}{x + 2}$

12. $$\frac{4x + 3}{10x} + \frac{1}{2x} - \frac{2}{5}$$
$$= \frac{4x + 3}{10x} + \frac{5 \cdot 1}{5 \cdot 2x} - \frac{2x \cdot 2}{2x \cdot 5}$$
$$= \frac{4x + 3}{10x} + \frac{5}{10x} - \frac{4x}{10x}$$
$$= \frac{4x + 3 + 5 - 4x}{10x}$$
$$= \frac{8}{10x}$$
$$= \frac{4}{5x}$$

Answer: **(C)** $\frac{4}{5x}$

13. $$\frac{x}{x + 7} \cdot (2x + 14) = \frac{x}{\cancel{x + 7}_{1}} \cdot \frac{2\cancel{(x + 7)}^{1}}{1}$$
$$= \frac{2x}{1}$$
$$= 2x$$

Answer: **(A)** $2x$

14. $$\frac{3x}{5y} \div \frac{x}{y} = \frac{3\cancel{x}^{1}}{5\cancel{y}_{1}} \cdot \frac{\cancel{y}^{1}}{\cancel{x}_{1}} = \frac{3}{5}$$

Answer: **(D)** $\frac{3}{5}$

15. $$\frac{20\sqrt{3}}{\sqrt{5}} = \frac{20\sqrt{3}}{\sqrt{5}} \cdot \frac{\sqrt{5}}{\sqrt{5}}$$
$$= \frac{20\sqrt{15}}{5}$$
$$= 4\sqrt{15}$$

Answer: **(B)** $4\sqrt{15}$

EQUATIONS, INEQUALITIES AND PROBLEMS IN ALGEBRA

Equations

1. a. An **equation** states that two quantities are equal.

   b. The solution to an equation is a number which can be substituted for the letter, or **variable**, to give a true statement.

   Example: In the equation $x + 7 = 10$,
   if 5 is substituted for x, the equation becomes $5 + 7 = 10$, which is false. If 3 is substituted for x, the equation becomes $3 + 7 = 10$, which is true. Therefore, $x = 3$ is a solution for the equation $x + 7 = 10$.

   c. To **solve an equation** means to find all solutions for the variables.

2. a. An equation has been solved when it is transformed or rearranged so that a variable is isolated on one side of the equal sign and a number is on the other side.

   b. There are two basic principles which are used to transform equations:

      I) The same quantity may be added to, or subtracted from, both sides of an equation.

      Example: To solve the equation $x - 3 = 2$, add 3 to both sides:

$$\begin{array}{rcr} x - 3 & = & 2 \\ + 3 & & +3 \\ \hline x & = & 5 \end{array}$$

      Adding 3 isolates x on one side and leaves a number on the other side. The solution to the equation is $x = 5$.

      Example: To solve the equation $y + 4 = 10$, subtract 4 from both sides (adding -4 to both sides will have the same effect):

$$\begin{array}{rcr} y + 4 & = & 10 \\ - 4 & & -4 \\ \hline y & = & 6 \end{array}$$

      The variable has been isolated on one side of the equation. The solution is $y = 6$.

II) Both sides of an equation may be multiplied by, or divided by, the same quantity.

Example: To solve $2a = 12$, divide both sides by 2:

$$\frac{2a}{2} = \frac{12}{2}$$
$$a = 6$$

Example: To solve $\frac{b}{5} = 10$, multiply both sides by 5:

$$5 \cdot \frac{b}{5} = 10 \cdot 5$$
$$b = 50$$

3. To solve equations containing more than one operation:

   a. First eliminate any number that is being added to or subtracted from the variable.

   b. Then eliminate any number that is multiplying or dividing the variable.

Illustration: Solve

$3x - 6 = 9$		
$+6 \quad +6$	Adding 6 eliminates -6.
$3x = 15$	
$\frac{3x}{3} = \frac{15}{3}$	Dividing by 3 eliminates the 3 which is multiplying the x.
$x = 5$	The solution to the original equation is $x = 5$.

4. A variable term may be added to, or subtracted from, both sides of an equation. This is necessary when the variable appears on both sides of the original equation.

Illustration: Solve

$6y + 9 = 2y + 1$ $-2y \quad -2y$	Eliminate the y term from the right side by subtracting $2y$ from both sides.
$4y + 9 = +1$ $-9 \quad -9$	Eliminate 9 from the left side by subtracting 9 from both sides.
$4y = -8$	
$\frac{4y}{4} = \frac{-8}{4}$	Divide both sides by 4 to eliminate the multiplication by 4 and isolate the y.
$y = -2$	

5. It may be necessary to first **simplify** the expression on each side of an equation by removing parentheses or combining like terms.

Illustration: Solve

$5z - 3(z - 2) = 8$	
$5z - 3z + 6 = 8$	Remove parentheses first.
$2z + 6 = 8$	Combine like terms.
$-6 \quad -6$	Subtract 6 from both sides.
$\frac{2z}{2} = \frac{2}{2}$	Divide by 2 to isolate the z.
$z = 1$	

6. To check the solution to any equation, replace the variable with the solution in the original equation, perform the indicated operations, and determine whether a true statement results.

Example: Earlier it was found that $x = 5$ is the solution for the equation $3x - 6 = 9$. To check, substitute 5 for x in the equation:

$3 \cdot 5 - 6 = 9$	Perform the operations on the left side.
$15 - 6 = 9$	
$9 = 9$	A true statement results; therefore the solution is correct.

Solving Problems

7. Many types of problems can be solved by using algebra. To solve a problem:

   a. Read it carefully. Determine what information is given and what information is unknown and must be found.

   b. Represent the *unknown* quantity with a letter.

   c. Write an equation that expresses the relationship given in the problem.

   d. Solve the equation.

Example: If 7 is added to twice a number, the result is 23. Find the number.

SOLUTION: Let x = the unknown number. Then write the equation:

$$\begin{aligned} 7 + 2x &= 23 \\ -7 \quad &\quad -7 \\ \hline \frac{2x}{2} &= \frac{16}{2} \\ x &= 8 \end{aligned}$$

Answer: 8

Example: There are 6 more women than men in a group of 26 people. How many women are there?

SOLUTION: Let m = the number of men. Then, m + 6 = the number of women.

$$\begin{aligned} (m + 6) + m &= 26 \\ m + 6 + m &= 26 \quad \text{Remove parentheses.} \\ 2m + 6 &= 26 \quad \text{Combine like terms.} \\ -6 \quad &\quad -6 \\ \hline \frac{2m}{2} &= \frac{20}{2} \\ m &= 10 \\ m + 6 &= 16 \end{aligned}$$

Answer: There are 16 women.

Example: John is 3 years older than Mary. If the sum of their ages is 39, how old is Mary?

SOLUTION: Let m = Mary's age. Then, $m + 3$ = John's age. The sum of their ages is 39.

$$\begin{aligned} m + (m + 3) &= 39 \\ m + m + 3 &= 39 \\ 2m + 3 &= 39 \\ -3 \quad & \quad -3 \\ \frac{2m}{2} &= \frac{36}{2} \\ m &= 18 \end{aligned}$$

Answer: Mary is 18 years old.

Consecutive Integer Problems

8. a. **Consecutive integers** are integers that follow one another.

Example: 7, 8, 9, and 10 are consecutive integers.
−5, −4, −3, −2 and −1 are consecutive integers.

b. Consecutive integers may be represented in algebra as:

$$x,\ x + 1,\ x + 2,\ x + 3,\ \ldots$$

Example: Find three consecutive integers whose sum is 39.

SOLUTION: Let x = first consecutive integer. Then, $x + 1$ = second consecutive integer and $x + 2$ = third consecutive integer.

$$\begin{aligned} x + (x + 1) + (x + 2) &= 39 \\ x + x + 1 + x + 2 &= 39 \\ 3x + 3 &= 39 \\ -3 \quad & \quad -3 \\ \frac{3x}{3} &= \frac{36}{3} \\ x &= 12 \end{aligned}$$

Answer: The integers are 12, 13 and 14.

9. Consecutive even and consecutive odd integers are both represented as x, $x + 2$, $x + 4$, $x + 6$, . . .

If x is even, then $x + 2$, $x + 4$, $x + 6$, . . . will all be even.
If x is odd, then $x + 2$, $x + 4$, $x + 6$, . . . will all be odd.

Example: Find four consecutive odd integers such that the sum of the largest and twice the smallest is 21.

SOLUTION: Let x, $x + 2$, $x + 4$, and $x + 6$ be the four consecutive odd integers. Here, x is the smallest and $x + 6$ is the largest. The largest integer plus twice the smallest is 21.

$$\begin{aligned} x + 6 + 2x &= 21 \\ 3x + 6 &= 21 \\ -6 \quad & \quad -6 \\ \frac{3x}{3} &= \frac{15}{3} \\ x &= 5 \end{aligned}$$

Answer: The integers are 5, 7, 9, and 11.

Motion Problems

10. **Motion problems** are based on the following relationship:

Rate · Time = Distance

Rate is usually given in miles per hour. Time is usually given in hours and distance is given in miles.

Example: A man traveled 225 miles in 5 hours. How fast was he traveling (what was his rate)?

SOLUTION: Let r = rate

$$\text{rate} \cdot \text{time} = \text{distance}$$
$$r \cdot 5 = 225$$
$$\frac{5r}{5} = \frac{225}{5}$$
$$r = 45 \text{ miles per hour}$$

Example: John and Henry start at the same time from cities 180 miles apart and travel toward each other. John travels at 40 miles per hour and Henry travels at 50 miles per hour. In how many hours will they meet?

SOLUTION: Let h = number of hours. Then, 40h = distance traveled by John, and 50h = distance traveled by Henry. The total distance is 180 miles.

$$40h + 50h = 180$$
$$\frac{90h}{90} = \frac{180}{90}$$
$$h = 2 \text{ hours}$$

Answer: They will meet in 2 hours.

Perimeter Problems

11. To solve a perimeter problem, express each side of the figure algebraically. The **perimeter** of the figure is equal to the sum of all of the sides.

Example: A rectangle has four sides. One side is the length and the side next to it is the width. The opposite sides of a rectangle are equal. In a particular rectangle, the length is one less than twice the width. If the perimeter is 16, find the length and the width.

SOLUTION:

Let w = width

Then 2w − 1 = length

The sum of the four sides is 16.

2w − 1

w w

2w − 1

$$w + (2w - 1) + w + (2w - 1) = 16$$
$$w + 2w - 1 + w + 2w - 1 = 16$$
$$6w - 2 = 16$$
$$+2 \quad +2$$
$$\frac{6w}{6} = \frac{18}{6}$$
$$w = 3$$
$$2w - 1 = 2(3) - 1 = 5$$

Answer: The width is 3 and the length is 5.

Ratio and Proportion Problems

12. a. A ratio is the quotient of two numbers. The ratio of 2 to 5 may be expressed $2 \div 5$, $\frac{2}{5}$, 2 is to 5, 2:5, or algebraically as $2x:5x$.

The numbers in a ratio are called the terms of the ratio.

Example: Two numbers are in the ratio 3:4. Their sum is 35. Find the numbers.

SOLUTION: Let $3x$ = the first number
$4x$ = the second number

Note that $\frac{3x}{4x} = \frac{3}{4} = 3:4$

The sum of the numbers is 35.

$$3x + 4x = 35$$
$$\frac{7x}{7} = \frac{35}{7}$$
$$x = 5$$
$$3x = 15$$
$$4x = 20$$

Answer: The numbers are 15 and 20.

b. A ratio involving more than two numbers may also be expressed algebraically. The ratio 2:3:7 is equal to $2x:3x:7x$. The individual quantities in the ratio are $2x$, $3x$, and $7x$.

13. a. A **proportion** states that two ratios are equal.

b. In the proportion $a:b = c:d$ (which may also be written $\frac{a}{b} = \frac{c}{d}$), the inner terms, b and c, are called the **means**; the outer terms, a and d, are called the **extremes**.

Example: In 3:6 = 5:10, the means are 6 and 5; the extremes are 3 and 10.

c. In any proportion, the product of the means equals the product of the extremes. In $a:b = c:d$, $bc = ad$.

Example: In 3:6 = 5:10, or $\frac{3}{6} = \frac{5}{10}$, $6 \cdot 5 = 3 \cdot 10$.

d. In many problems, the quantities involved are in proportion. If three quantities are given in a problem and the fourth quantity is unknown, determine whether the quantities should form a proportion. The proportion will be the equation for the problem.

Example: A tree that is 20 feet tall casts a shadow 12 feet long. At the same time, a pole casts a shadow 3 feet long. How tall is the pole?

SOLUTION: Let p = height of pole. The heights of objects and their shadows are in proportion.

$$\frac{\text{tree}}{\text{tree's shadow}} = \frac{\text{pole}}{\text{pole's shadow}}$$

$$\frac{20}{12} = \frac{p}{3}$$

$$12p = 60$$ The product of the means equals the product of the extremes.

$$\frac{12p}{12} = \frac{60}{12}$$

$$p = 5$$

Answer: The pole is 5 feet tall.

Example: The scale on a map is 3 cm = 500 km. If two cities appear 15 cm apart on the map, how far apart are they actually?

SOLUTION: Let d = actual distance. The quantities on maps and scale drawings are in proportion with the quantities they represent.

$$\frac{\text{first map distance}}{\text{first actual distance}} = \frac{\text{second map distance}}{\text{second actual distance}}$$

$$\frac{3\text{ cm}}{500\text{ km}} = \frac{15\text{ cm}}{d\text{ km}}$$

$$3d = 7500$$ The product of the means equals the product of the extremes.

$$\frac{3d}{3} = \frac{7500}{3}$$

$$d = 2500$$

Answer: The cities are 2500 km apart.

Percent Problems

14. **Percent** problems may be solved algebraically by translating the relationship in the problem into an equation. The word "of" means multiplication, and "is" means equal to.

Example: 45% of what number is 27?

SOLUTION: Let n = the unknown number. 45% of n is 27.

$$.45n = 27$$ Change the % to a decimal (45% = .45)

$$45n = 2700$$ Multiplying both sides by 100 eliminates the decimal.

$$\frac{45n}{45} = \frac{2700}{45}$$

$$n = 60$$

Example: Mr. Jones receives a salary raise from $15,000 to $16,200. Find the percent of increase.

SOLUTION: Let p = percent. The increase is 16,200 − 15,000 = 1,200. What percent of 15,000 is 1,200?

$$p \cdot 15{,}000 = 1{,}200$$

$$\frac{15000p}{15000} = \frac{1200}{15000}$$

$$p = .08$$

$$p = 8\%$$

15. **Interest** is the price paid for the use of money in loans, savings and investments. Interest problems are solved using the formula **I = prt**, where:

I = interest
p = principal (amount of money bearing interest)
r = rate of interest, in %
t = time, in years

Example: How long must $2000 be invested at 6% to earn $240 in interest?

SOLUTION:

Let t = time
I = $240
p = $2000
r = 6% or .06

$$240 = 2000(.06)t$$
$$\frac{240}{120} = \frac{120t}{120}$$
$$2 = t$$

Answer: The $2000 must be invested for 2 years.

16. a. A **discount** is a percent that is deducted from a marked price. The marked price is considered to be 100% of itself.

Example: If an item is discounted 20%, its selling price is 100% − 20%, or 80%, of its marked price.

Example: A radio is tagged with a sale price of $42.50, which is 15% off the regular price. What is the regular price?

SOLUTION: Let r = regular price. The sale price is 100% − 15%, or 85%, of the regular price. 85% of r = $42.50

$$.85r = \$42.50$$
$$\frac{85r}{85} = \frac{4250}{85} \quad \text{Multiply by 100 to eliminate the decimals.}$$
$$r = 50$$

Answer: The regular price was $50.

b. If two discounts are given in a problem, an intermediate price is computed by taking the first discount on the marked price. The second discount is then computed on the intermediate price.

Example: An appliance company gives a 15% discount for purchases made during a sale, and an additional 5% discount if payment is made in cash. What will the price of a $800 refrigerator be if both discounts are taken?

SOLUTION: First discount: 100% − 15% = 85%

After the first discount, the refrigerator will cost:

85% of $800 = .85($800)
= $680

The intermediate price is $680.

Second discount: 100% − 5% = 95%.

After the second discount, the refrigerator will cost:

95% of $680 = .95($680)
= $646

Answer: The final price will be $646.

17. a. **Profit** is the amount of money added to the dealer's cost of an item to find the selling price. The cost price is considered 100% of itself.

Example: If the profit is 20% of the cost, the selling price must be 100% + 20%, or 120% of the cost.

Example: A furniture dealer sells a sofa at \$870, which represents a 45% profit over the cost. What was the cost to the dealer?

SOLUTION: Let c = cost price. 100% + 45% = 145%. The selling price is 145% of the cost.

$$\begin{aligned} 145\% \text{ of } c &= \$870 \\ 1.45c &= 870 \\ \frac{145c}{145} &= \frac{87000}{145} \\ c &= 600 \end{aligned}$$

Answer: The sofa cost the dealer \$600.

b. If an article is sold at a **loss**, the amount of the loss is deducted from the cost price to find the selling price.

Example: An article that is sold at a 25% loss has a selling price of 100% − 25%, or 75%, of the cost price.

Example: Mr. Charles bought a car for \$8000. After a while he sold it to Mr. David at a 30% loss. What did Mr. David pay for the car?

SOLUTION: The car was sold for 100% − 30%, or 70%, of its cost price.

$$\begin{aligned} 70\% \text{ of } \$8000 &= .70(\$8000) \\ &= \$5600 \end{aligned}$$

Answer: Mr. David paid \$5600 for the car.

18. **Tax** is computed by finding a percent of a base amount.

Example: A homeowner pays \$2500 in school taxes. What is the assessed value of his property if school taxes are 3.2% of the assessed value?

SOLUTION: Let v = assessed value.

$$\begin{aligned} 3.2\% \text{ of } v &= 2500 \\ .032v &= 2500 \\ \frac{32v}{32} &= \frac{2500000}{32} \quad \text{(Multiply by 1000 to eliminate decimals)} \\ v &= 78125 \end{aligned}$$

Answer: The value of the property is \$78,125.

Inequalities

19. a. The = symbol indicates the relationship between two equal quantities. The symbols used to indicate other relationships between two quantities are:

$\neq$ not equal to
$>$ greater than
$<$ less than
$\geqslant$ greater than or equal to
$\leqslant$ less than or equal to

b. A number is **greater** than any number appearing to its left on the number line. A number is **less** than any number appearing to its right on the number line.

Examples:

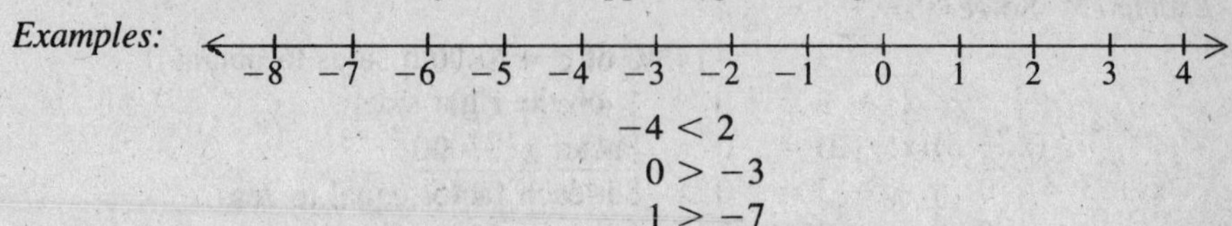

$$-4 < 2$$
$$0 > -3$$
$$1 > -7$$

20. a. An **inequality** states that one quantity is greater than, or less than, another quantity.

b. Inequalities are solved in the same way as equations, except that in multiplying or dividing both sides of an inequality by a negative quantity, the inequality symbol is reversed.

Example: Solve for x:

$3x - 4 > 11$

$\quad + 4 \quad +4$ Add 4 to both sides.

$\frac{3x}{3} > \frac{15}{3}$ Divide both sides by 3. Since 3 is positive, the inequality symbol remains the same.

$x > 5$

The solution $x > 5$ means that all numbers greater than 5 are solutions to the inequality.

Example: Solve for y:

$2y + 3 > 7y - 2$

$-7y \quad -7y$ Subtract 7y from both sides.

$-5y + 3 > -2$

$\quad - 3 \quad - 3$ Subtract 3 from both sides.

$-5y > -5$ Divide both sides by −5. When dividing both sides by a negative number, reverse the inequality symbol.

$y < 1$

Quadratic Equations

21. a. A **quadratic equation** is an equation in which the variable has 2 as its greatest exponent. Quadratic equations may be put into the form $ax^2 + bx + c = 0$, where a, b and c are constants and $a \neq 0$.

b. The solution of quadratic equations is based on the principle that if the product of two quantities is zero, at least one of those quantities must be zero.

If one side of a quadratic equation is zero and the other side can be written as the product of two factors, each of those factors may be set equal to zero and the resulting equations solved.

Example: Solve $x^2 - 7x + 10 = 0$

The factors of the trinomial are $(x - 2)(x - 5) = 0$

Set each factor equal to zero: $x - 2 = 0; \quad x - 5 = 0$

Solve each equation: $x = 2 \qquad x = 5$

The solutions of $x^2 - 7x + 10 = 0$ are 2 and 5.

Example: Solve $x^2 - 5 = 4$

$$\begin{array}{rl} x^2 - 5 = & 4 \\ -4 & -4 \\ \hline x^2 - 9 = & 0 \end{array}$$
Add -4 to both sides to obtain 0 on the right side.

$(x + 3)(x - 3) = 0$ Factor $x^2 - 9$.

$$\begin{array}{rl|rl} x + 3 = & 0 & x - 3 = & 0 \\ -3 & -3 & +3 & +3 \\ \hline x = & -3 & x = & 3 \end{array}$$
Set each factor equal to zero.
Solve each equation.

The solutions of $x^2 - 5 = 4$ are 3 and -3.

Example: Solve $3z^2 - 12z = 0$

$3z(z - 4) = 0$ Factor $3z^2 - 12z$.

$$\begin{array}{c|rl} \frac{3z}{3} = \frac{0}{3} & z - 4 = & 0 \\ & +4 & +4 \\ \hline z = 0 & z = & 4 \end{array}$$
Set each factor equal to zero.
Solve each equation.

The solutions of $3z^2 - 12z = 0$ are 0 and 4.

22. If a quadratic equation is written in the form $ax^2 + bx + c = 0$, where a, b and c are constants and $a \neq 0$, the equation may be solved by using the formula:

$$x = \frac{-b \pm \sqrt{b^2 - 4ac}}{2a}$$

Example: Solve $2x^2 - 7x - 15 = 0$

$a = 2,\ b = -7,\ c = -15$

$$x = \frac{-(-7) \pm \sqrt{(-7)^2 - 4(2)(-15)}}{2(2)}$$

$$= \frac{7 \pm \sqrt{49 + 120}}{4}$$

$$= \frac{7 \pm \sqrt{169}}{4}$$

$$= \frac{7 \pm 13}{4}$$

$$x = \frac{7 + 13}{4} = \frac{20}{4} = 5 \qquad x = \frac{7 - 13}{4} = \frac{-6}{4} = -1\frac{1}{2}$$

The solutions of $2x^2 - 7x - 15 = 0$ are 5 and $-1\frac{1}{2}$.

Simultaneous Equations

23. If two equations in two unknowns are considered at the same time, their simultaneous solution is the pair of values which satisfies both equations.

Example: The simultaneous equations

$$2x + y = 10$$
$$x + 3y = 15$$

have the pair $x = 3$, $y = 4$ as a common solution, since

$$2 \cdot 3 + 4 = 10$$
and $$3 + 3 \cdot 4 = 15$$

24. Simultaneous equations are solved by eliminating one unknown, leaving a new equation in the remaining unknown. The new equation is solved and its solution substituted in one of the original equations to find the other unknown.

25. Addition and subtraction methods for eliminating an unknown:
For these methods, the equations must be arranged so that the equal signs are in a column and the like unknowns are in columns.

a. Addition: Add the two equations together if doing so eliminates one unknown.

Example:

$$\begin{aligned} x + y &= 10 \\ \underline{x - y} &= \underline{4} \\ 2x &= 14 \\ x &= 7 \end{aligned}$$

Adding the equations eliminates the y

Solve for x

Substitute $x = 7$ in one original equation and solve for y:

$$\begin{aligned} 7 + y &= 10 \\ y &= 3 \\ x = 7, \quad y &= 3 \end{aligned}$$

b. Subtraction: Subtract one equation from the other if doing so eliminates one unknown.

Example:

$$\begin{aligned} 5a + 2b &= 13 \\ \underline{3a + 2b} &= \underline{11} \\ 2a &= 2 \\ a &= 1 \end{aligned}$$

Subtracting eliminates b

Solve for a

Substitute $a = 1$ in one original equation and solve for b:

$$\begin{aligned} 3 \cdot 1 + 2b &= 11 \\ 3 + 2b &= 11 \\ 2b &= 8 \\ b &= 4 \\ a = 1, \quad b &= 4 \end{aligned}$$

c. Multiplication with addition or subtraction: Adding or subtracting will eliminate an unknown only if that unknown had the same coefficient in both equations. It may be necessary to multiply both sides of one or both equations by some number in order to have the same coefficient for an unknown.

Example:

$$2p + q = 4$$
$$3p - 2q = 13$$

Multiply both sides of the first equation by 2, then add the equations:

$$2(2p + q) = 2(4) \qquad 4p + 2q = 8$$
$$3p - 2q = 13 \qquad \underline{3p - 2q = 13}$$
$$7p = 21$$
$$p = 3$$

Substituting $p = 3$ in $2p + q = 4$,

$$2 \cdot 3 + q = 4$$
$$6 + q = 4$$
$$q = -2$$
$$p = 3, \quad q = -2$$

Example:

$$6r - 2s = 10$$
$$5r + 3s = 13$$

Multiplying both sides of the first equation by 3 and both sides of the second equation by 2 will yield 6s in each equation. The equations may then be added.

$$3(6r - 2s) = 3(10) \qquad 18r - 6s = 30$$
$$2(5r + 3s) = 2(13) \qquad \underline{10r + 6s = 26}$$
$$28r = 56$$
$$r = 2$$

Substituting $r = 2$ in $5r + 3s = 13$,

$$5 \cdot 2 + 3s = 13$$
$$10 + 3s = 13$$
$$3s = 3$$
$$s = 1$$
$$r = 2, \quad s = 1$$

26. Substitution method for eliminating an unknown:
The substitution method may be used if, in one equation, one unknown is expressed in terms of the other.

Example:

$$3x + 4y = 18$$
$$y = x + 1$$

Substitute $x + 1$ for y in the first equation:

$$3x + 4(x + 1) = 18$$
$$3x + 4x + 4 = 18$$
$$7x + 4 = 18$$
$$7x = 14$$
$$x = 2$$

$$y = x + 1$$
$$y = 2 + 1$$
$$y = 3$$
$$x = 2, \quad y = 3$$

Equations Containing Radicals (Square Roots)

27. a. To eliminate a radical sign in an equation, isolate the radical on one side of the equation, then square both sides.

Example: Solve $\sqrt{5x - 1} - 3 = 0$

$\sqrt{5x - 1} = 3$ Isolate the radical

$(\sqrt{5x - 1})^2 = (3)^2$ Square each side

$5x - 1 = 9$

$5x = 10$

$x = 2$

b. Sometimes an extraneous solution (that is, a value which is not a solution to the original equation) will result from squaring both sides of the equation, even though the proper method has been followed. To be sure the solution is not extraneous, check by substituting it in the original equation.

Example: $\sqrt{x^2 + 7} = x - 1$

$(\sqrt{x^2 + 7})^2 = (x - 1)^2$

$x^2 + 7 = x^2 - 2x + 1$

$7 = -2x + 1$

$6 = -2x$

$-3 = x$

Check $x = -3$:

Does $\sqrt{(-3)^2 + 7} = -3 - 1$?

$\sqrt{9 + 7} = -4$?

$\sqrt{16} = -4$?

Since the $\sqrt{\ }$ symbol refers to the positive square root only, $\sqrt{16} \neq -4$, and $x = -3$ is an extraneous solution.

Practice Problems — Algebra

1. If $6x - (2x + 6) = x + 3$, then $x =$
   (A) -3 (C) 1
   (B) -1 (D) 3

2. If $y^2 - 5y - 6 = 0$, then $y =$
   (A) 6 or -1 (C) -2 or 3
   (B) -6 or 1 (D) 2 or -3

3. Solve for z: $8z + 5 - 10z > -3$
   (A) $z > 4$ (C) $z < 4$
   (B) $z > -4$ (D) $z < -4$

4. If $2x^3 + 5x = 4x^3 - 2x^3 + 10$, then $x =$
   (A) -2 (C) 1
   (B) -1 (D) 2

5. One number is three times another number. If their difference is 30, the smaller number is
   (A) 5 (C) 15
   (B) 10 (D) 20

6. The perimeter of the figure below is 41. The length of the longest side is
   (A) 10 (C) 12
   (B) 11 (D) 13

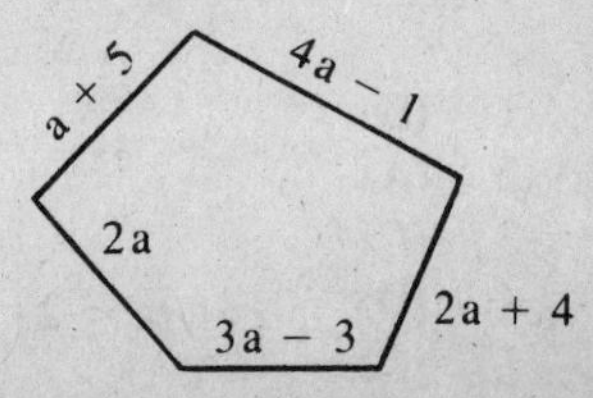

7. The sum of four consecutive even integers is 12. The smallest of the integers is
   (A) -2 (C) 2
   (B) 0 (D) 4

8. An estate was divided among three heirs, A, B and C, in the ratio 2:3:4. If the total estate was $22,500, what was the smallest inheritance?
   (A) $1000 (C) $2500
   (B) $1250 (D) $5000

9. A dealer buys a TV set for $550 and wishes to sell it at a 20% profit. What should his selling price be?
   (A) $570 (C) $660
   (B) $600 (D) $672

10. Michael earns $50 for 8 hours of work. At the same rate of pay, how much will he earn for 28 hours of work?
    (A) $150 (C) $186
    (B) $175 (D) $232

11. Mrs. Smith wishes to purchase a freezer with a list price of $500. If she waits for a "15% off" sale and receives an additional discount of 2% for paying cash, how much will she save?
    (A) $75.50 (C) $85.00
    (B) $83.50 (D) $150.00

12. A photograph is 8″ wide and 10″ long. If it is enlarged so that the new length is 25″, the new width will be
    (A) $18\frac{1}{2}$″ (C) 24″
    (B) 20″ (D) $31\frac{1}{4}$″

13. Jean sells cosmetics, earning a 12% commission on all sales. How much will she need in sales to earn $300 in commission?
    (A) $1800 (C) $3600
    (B) $2500 (D) $4000

14. Mr. Taylor leaves home at 8 AM, traveling at 45 miles per hour. Mrs. Taylor follows him, leaving home at 10 AM and traveling at 55 miles per hour. How long will it take Mrs. Taylor to catch up with Mr. Taylor?
    (A) 7 hours (C) 9 hours
    (B) 8 hours (D) 10 hours

15. Sam buys a jacket marked $85. He pays $90.95 including sales tax. What percent sales tax does he pay?
    (A) 4% (C) 6%
    (B) 5% (D) 7%

16. If one side of a square is decreased by 1 inch and an adjacent side is increased by 4 inches, a rectangle is formed whose area is 36 square inches. The number of inches in the side of the original square is
    (A) 4 (C) 6
    (B) 5 (D) 8

17. If $x^2 - x = 1$, $x =$
    (A) ± 1 (C) $\frac{1 \pm \sqrt{5}}{2}$
    (B) $\frac{1 \pm \sqrt{3}}{2}$ (D) $\frac{-1 \pm \sqrt{5}}{2}$

18. If $x + \sqrt{x - 2} = 2$, then $x =$
    (A) 2 or 3 (C) -3 only
    (B) 2 only (D) -2 or -3

19. If $x + y = 9$ and $y = x - 3$, then $y =$
    (A) 2 (C) 3
    (B) -2 (D) -3

20. Solve for x:
    $$2x + 3y = 12$$
    $$3x - y = 7$$
    (A) 1 (C) 3
    (B) 2 (D) 4½

Practice Problems — Correct Answers

1. **(D)**	6. **(B)**	11. **(B)**	16. **(B)**
2. **(A)**	7. **(B)**	12. **(B)**	17. **(C)**
3. **(C)**	8. **(D)**	13. **(B)**	18. **(B)**
4. **(D)**	9. **(C)**	14. **(C)**	19. **(C)**
5. **(C)**	10. **(B)**	15. **(D)**	20. **(C)**

Problem Solutions — Algebra

1. $$6x - (2x + 6) = x + 3$$

$$6x - 2x - 6 = x + 3$$ Remove parentheses first.

$$4x - 6 = x + 3$$ Combine the like terms on the left side.

$$\underline{-x} \qquad \underline{-x}$$ Eliminate the x term from the right side.

$$3x - 6 = 3$$

$$\underline{+6} \qquad \underline{+6}$$ Eliminate the number term from the left side.

$$\frac{3x}{3} = \frac{9}{3}$$ Divide both sides by 3 to isolate x.

$$x = 3$$

Answer: **(D)** 3

2. $$y^2 - 5y - 6 = 0$$

$$(y - 6)(y + 1) = 0$$ Factor the trinomial side of the quadratic equation.

$$y - 6 = 0 \quad \Big| \quad y + 1 = 0$$ Set each factor equal to zero.

$$\underline{+6} \quad \underline{+6} \quad \Big| \quad \underline{-1} \quad \underline{-1}$$ Solve each equation.

$$y = 6 \quad \Big| \quad y = -1$$

Answer: **(A)** 6 or -1

3. $$8z + 5 - 10z > -3$$

$$-2z + 5 > -3$$ Combine the like terms on the left side.

$$\underline{-5} \quad \underline{-5}$$

$$-2z > -8$$ Divide both sides by -2 and reverse the inequality symbol.

$$z < 4$$

Answer: **(C)** $z < 4$

4. $$2x^3 + 5x = 4x^3 - 2x^3 + 10$$

$$2x^3 + 5x = 2x^3 + 10$$ Combine the like terms on the right side.

$$\underline{-2x^3} \qquad \underline{-2x^3}$$ Subtracting $2x^3$ from both sides leaves a simple equation.

$$\frac{5x}{5} = \frac{10}{5}$$

$$x = 2$$

Answer: **(D)** 2

5. Let n = the smaller number. Then $3n$ = the larger number. The difference of the numbers is 30.

$$3n - n = 30$$

$$\frac{2n}{2} = \frac{30}{2}$$

$$n = 15$$

Answer: **(C)** 15

6. The perimeter is equal to the sum of the sides.

$$a + 5 + 4a - 1 + 2a + 4 + 3a - 3 + 2a = 41$$

$$12a + 5 = 41 \quad \text{Combine like terms.}$$

$$\underline{\;\;-5 \quad -5}$$

$$\frac{12a}{12} = \frac{36}{12}$$

$$a = 3$$

The sides are:

$$a + 5 = 3 + 5 = 8$$
$$4a - 1 = 4 \cdot 3 - 1 = 11$$
$$2a + 4 = 2 \cdot 3 + 4 = 10$$
$$3a - 3 = 3 \cdot 3 - 3 = 6$$
$$2a = 2 \cdot 3 = 6$$

The longest side is 11.

Answer: **(B)** 11

7. Let x, x + 2, x + 4 and x + 6 represent the four consecutive even integers. The sum of the integers is 12.

$$x + x + 2 + x + 4 + x + 6 = 12$$
$$4x + 12 = 12$$
$$\underline{\;\;-12 \quad -12}$$
$$\frac{4x}{4} = \frac{0}{4}$$
$$x = 0$$
$$x + 2 = 2$$
$$x + 4 = 4$$
$$x + 6 = 6$$

The smallest integer is 0.

Answer: **(B)** 0

8. Let 2x, 3x and 4x represent the shares of the inheritance. The total estate was \$22,500.

$$2x + 3x + 4x = 22500$$
$$\frac{9x}{9} = \frac{22500}{9}$$
$$x = 2500$$
$$2x = 2 \cdot 2500 = 5000$$
$$3x = 3 \cdot 2500 = 7500$$
$$4x = 4 \cdot 2500 = 10{,}000$$

The smallest inheritance was 2x, or \$5000.

Answer: **(D)** \$5000

9. His selling price will be (100% + 20%) of his cost price.

$$120\% \text{ of } \$550 = 1.20(\$550)$$
$$= \$660$$

Answer: **(C)** \$660

10. The amount earned is proportional to the number of hours worked.

Let m = unknown pay

$$\frac{m}{28} = \frac{50}{8}$$

$$8m = 28 \cdot 50$$ The product of the means is equal to the product of the extremes.

$$\frac{8m}{8} = \frac{1400}{8}$$

$$m = 175$$

Answer: **(B)** $175

11. The selling price after the 15% discount is 85% of list.

$$\text{selling price} = .85(500) = 425$$

The selling price after the additional 2% discount is 98% of 425.

$$\text{new selling price} = .98(425) = 416.50$$

The original price was $500. Mrs. Smith buys at $416.50. She saves

$$\$500 - \$416.50 = \$83.50$$

Answer: **(B)** $83.50

12. The old dimensions and the new dimensions are in proportion. Let w = new width.

$$\frac{\text{new width}}{\text{old width}} = \frac{\text{new length}}{\text{old length}}$$

$$\frac{w}{8} = \frac{25}{10}$$

$$10w = 200$$

$$w = 20$$

Answer: **(B)** 20″

13. Let s = needed sales. 12% of sales will be $300.

$$\frac{.12s}{.12} = \frac{300}{.12}$$ Divide by .12, or first multiply by 100 to clear the decimal, then divide by 12.

$$s = 2500$$

Answer: **(B)** $2500

14. Let h = the number of hours needed by Mrs. Taylor. Mr. Taylor started two hours earlier; therefore, he travels h + 2 hours. Mrs. Taylor's distance is 55h. Mr. Taylor's distance is 45(h + 2). When Mrs. Taylor catches up with Mr. Taylor, they will have traveled equal distances.

$$55h = 45(h + 2)$$

$$55h = 45h + 90$$

$$-45h = -45h$$

$$10h = 90$$

$$h = 9$$

Answer: **(C)** 9 hours

15. The amount of tax is $90.95 − $85 = $5.95. Find the percent $5.95 is of $85.
Let p = percent.

$$\frac{p \cdot 85}{85} = \frac{5.95}{85}$$
$$p = .07 = 7\%$$

Answer: **(D)** 7%

16. Let x = side of original square
then $x - 1$ = width of new rectangle
$x + 4$ = length of new rectangle

$$(x - 1)(x + 4) = 36 \quad \text{The area of the rectangle is 36.}$$
$$x^2 + 3x - 4 = 36$$
$$-36 \quad -36$$
$$x^2 + 3x - 40 = 0$$
$$(x - 5)(x + 8) = 0$$
$$x - 5 = 0 \quad x + 8 = 0 \quad \text{Impossible to have a negative length}$$
$$x = 5 \quad x = -8$$

Answer: **(B)** 5

17. $x^2 - x = 1$ may be rewritten
$$1x^2 - 1x - 1 = 0$$
Using the quadratic formula with $a = 1$, $b = -1$, and $c = -1$,
$$x = \frac{-(-1) \pm \sqrt{(-1)^2 - 4(1)(-1)}}{2(1)}$$
$$= \frac{1 \pm \sqrt{1 + 4}}{2}$$
$$= \frac{1 \pm \sqrt{5}}{2}$$

Answer: **(C)** $\frac{1 \pm \sqrt{5}}{2}$

18.
$$x + \sqrt{x - 2} = 2$$
$$-x \qquad -x$$
$$\sqrt{x - 2} = 2 - x$$
$$(\sqrt{x - 2})^2 = (2 - x)^2$$
$$x - 2 = 4 - 4x + x^2$$
$$-x + 2 \qquad +2 - x$$
$$0 = 6 - 5x + x^2$$
$$x^2 - 5x + 6 = 0$$
$$(x - 3)(x - 2) = 0$$
$$x - 3 = 0 \quad x - 2 = 0$$
$$x = 3 \quad x = 2$$

Check

$x = 3$:
$x + \sqrt{3 - 2} = 2?$
$3 + \sqrt{1} = 2?$
$3 + 1 \neq 2$
$x = 3$ is extraneous

$x = 2$:
$2 + \sqrt{2 - 2} = 2?$
$2 + \sqrt{0} = 2?$
$2 + 0 \neq 2$
$x = 2$ checks

Answer: **(B)** 2 only

19. Substitute $y = x - 3$ in the first equation:

$$\begin{aligned} x + (x - 3) &= 9 \\ x + x - 3 &= 9 \\ 2x - 3 &= 9 \\ 2x &= 12 \\ x &= 6 \end{aligned}$$

$$\begin{aligned} y &= x - 3 \\ y &= 6 - 3 = 3 \end{aligned}$$

Answer: **(C)** 3

20. Multiply both sides of the second equation by 3, then add:

$$\begin{aligned} 2x + 3y = 12 &\longrightarrow 2x + 3y = 12 \\ 3(3x - y) = (7)3 &\longrightarrow \underline{9x - 3y = 21} \\ & \quad 11x = 33 \\ & \quad x = 3 \end{aligned}$$

Answer: **(C)** 3

GEOMETRY

Angles

1. a. An **angle** is the figure formed by two lines meeting at a point.

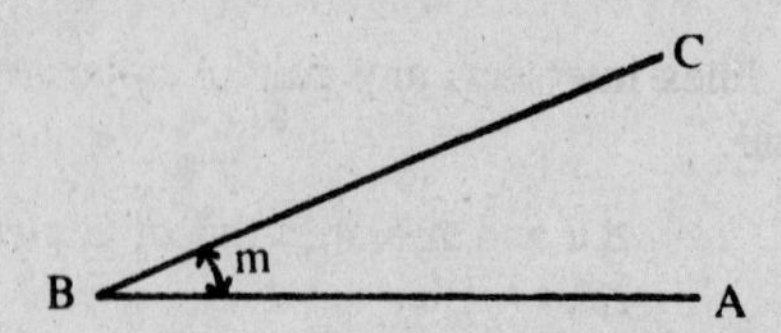

b. The point B is the **vertex** of the angle and the lines BA and BC are the **sides** of the angle.

2. There are three common ways of naming an angle:

a. By a small letter or figure written within the angle, as ∡m.

b. By the capital letter at its vertex, as ∡B.

c. By three capital letters, the middle letter being the vertex letter, as ∡ABC.

3. a. When two straight lines intersect (cut each other), four angles are formed. If these four angles are equal, each angle is a **right angle** and contains 90°. The symbol ⏋is used to indicate a right angle.

Example:

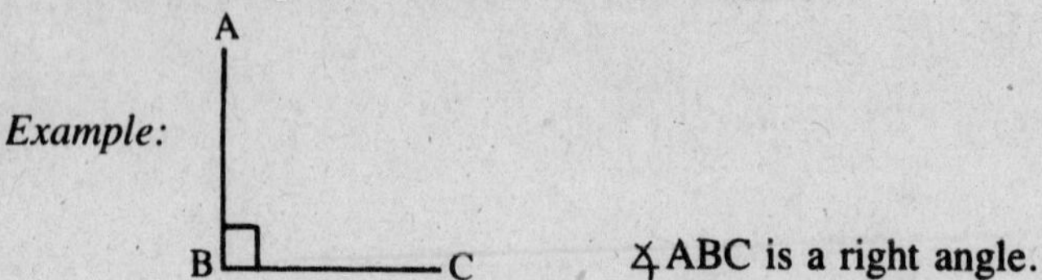

b. An angle less than a right angle is an **acute angle**.

c. If the two sides of an angle extend in opposite directions forming a straight line, the angle is a **straight angle** and contains 180°.

d. An angle greater than a right angle (90°) and less than a straight angle (180°) is an **obtuse angle**.

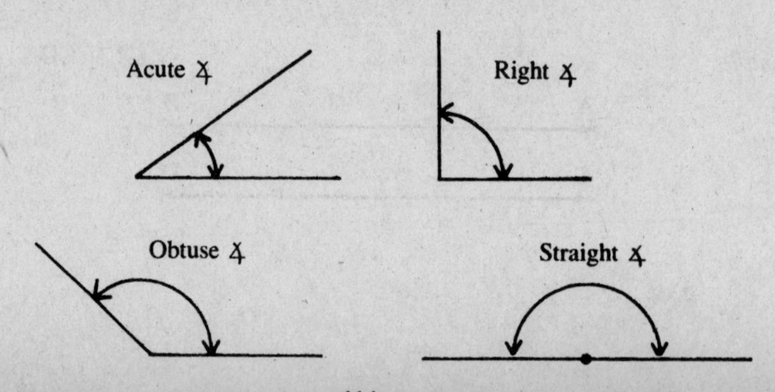

4. a. Two angles are **complementary** if their sum is 90°. To find the complement of an angle, subtract the given number of degrees from 90°.

Example: The complement of 60° = 90° − 60° = 30°.

b. Two angles are **supplementary** if their sum is 180°. To find the supplement of an angle, subtract the given number of degrees from 180°.

Example: The supplement of 60° = 180° − 60° = 120°.

5. When two straight lines intersect, any pair of opposite angles are called **vertical angles** and are equal.

$\measuredangle a$ and $\measuredangle b$ are vertical angles
$\measuredangle a = \measuredangle b$
$\measuredangle c$ and $\measuredangle d$ are vertical angles
$\measuredangle c = \measuredangle d$

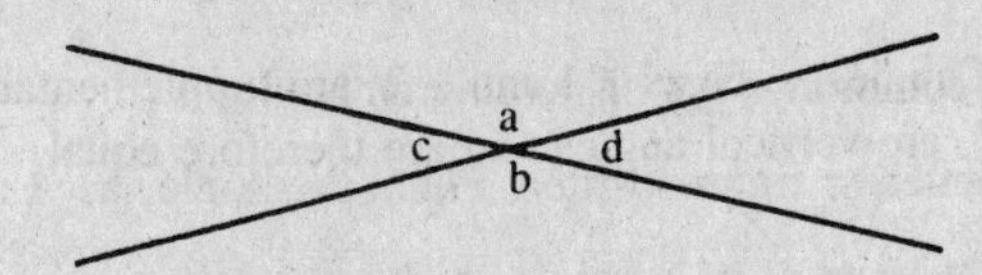

6. Two lines are **perpendicular** to each other if they meet to form a right angle. The symbol $\perp$ is used to indicate that the lines are perpendicular.

Example:

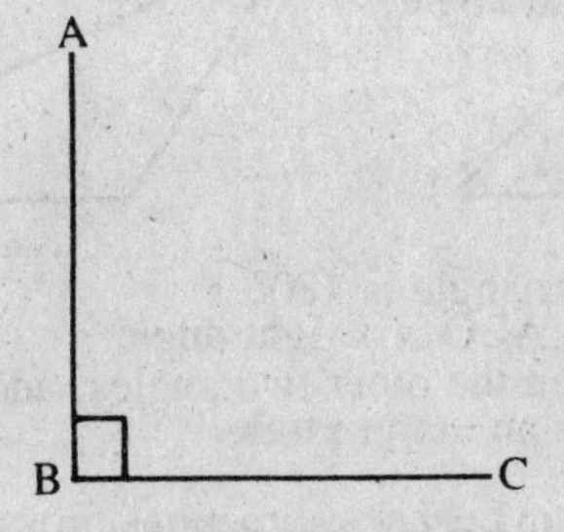

$\measuredangle ABC$ is a right angle.
Therefore, $AB \perp BC$.

7. a. Lines that do not meet no matter how far they are extended are called **parallel lines**. The symbol || is used to indicate that two lines are parallel.

Example: AB || CD

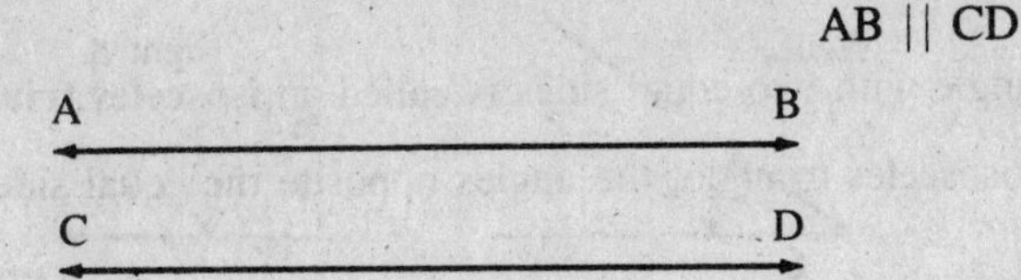

b. A line that intersects parallel lines is called a **transversal**. The pairs of angles formed have special names and relationships.

Example:

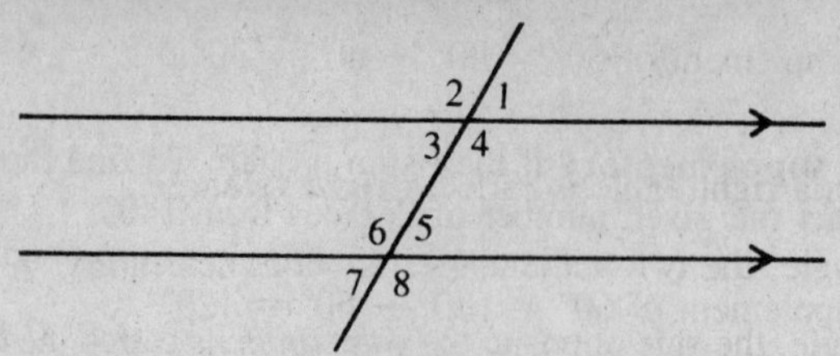

alternate interior angles:

$\measuredangle 3 = \measuredangle 5$
$\measuredangle 4 = \measuredangle 6$

corresponding angles:

$\measuredangle 1 = \measuredangle 5$
$\measuredangle 2 = \measuredangle 6$
$\measuredangle 3 = \measuredangle 7$
$\measuredangle 4 = \measuredangle 8$

Several pairs of angles, such as $\measuredangle 1$ and $\measuredangle 2$, are supplementary. Several pairs, such as $\measuredangle 6$ and $\measuredangle 8$, are vertical angles and are therefore equal.

Triangles

8. A triangle is a closed, three-sided figure. The following figures are triangles.

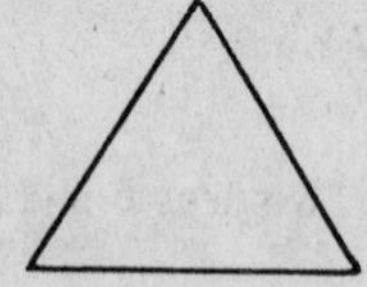
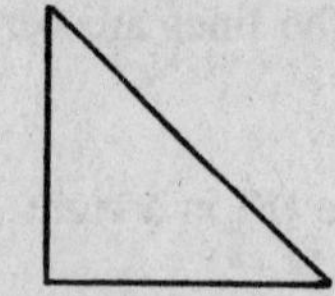
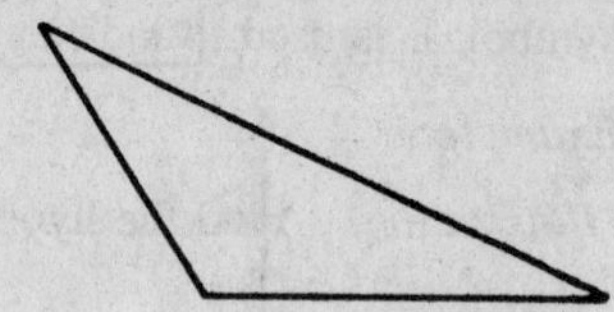

9. a. The sum of the three angles of a triangle is 180°.

   b. To find an angle of a triangle given the other two angles, add the given angles and subtract their sum from 180°.

Illustration: Two angles of a triangle are 60° and 40°. Find the third angle.

SOLUTION: $60° + 40° = 100°$
$180° - 100° = 80°$

Answer: The third angle is 80°.

10. a. A triangle with two equal sides is called an **isosceles triangle**.

    b. In an isosceles triangle, the angles opposite the equal sides are also equal.

Example:

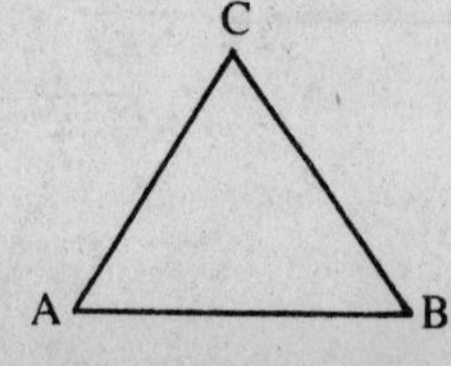

If AC = BC, then $\measuredangle A = \measuredangle B$

11. a. A triangle with all three sides equal is called an **equilateral triangle**.
    b. Each angle of an equilateral triangle is 60°.

12. a. A triangle with a right angle is called a **right triangle**.
    b. In a right triangle, the two acute angles are complementary.
    c. In a right triangle, the side opposite the right angle is called the **hypotenuse** and is the longest side. The other two sides are called **legs**.

Example:

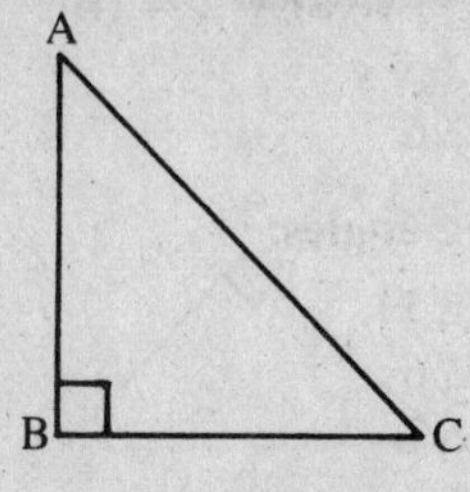

In right triangle ABC, AC is the hypotenuse. AB and BC are the legs.

13. The **Pythagorean Theorem** states that in a right triangle, the square of the hypotenuse equals the sum of the squares of the legs.

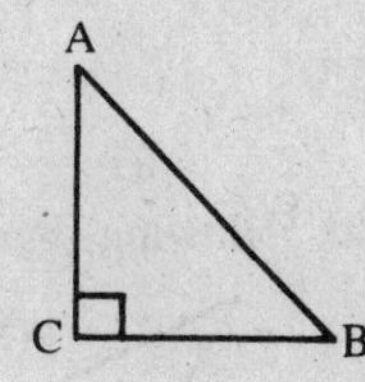

$$(AC)^2 + (BC)^2 = (AB)^2$$

Illustration: Find the hypotenuse (h) in a right triangle that has legs 6 and 8.

SOLUTION:

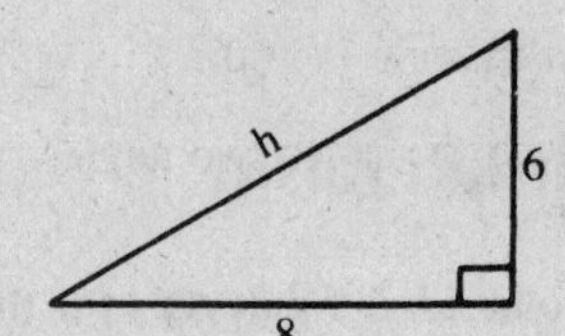

$$6^2 + 8^2 = h^2$$
$$36 + 64 = h^2$$
$$100 = h^2$$
$$\sqrt{100} = h$$
$$10 = h$$

Illustration: One leg of a right triangle is 5. The hypotenuse is 13. Find the other leg.

SOLUTION: Let the unknown leg be represented by x.

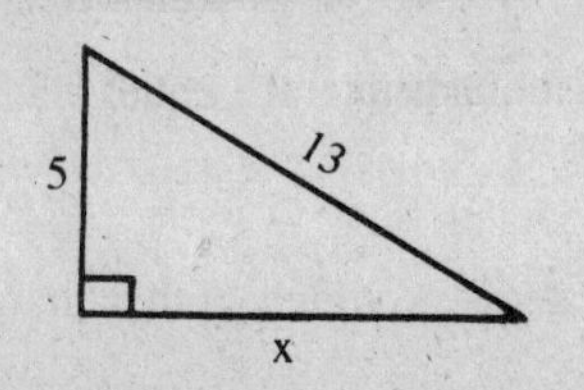

$$5^2 + x^2 = 13^2$$
$$25 + x^2 = 169$$
$$-25 \qquad -25$$
$$x^2 = 144$$
$$x = \sqrt{144}$$
$$x = 12$$

Answer: The other leg is 12.

14. a. In a right triangle with equal legs (an isosceles right triangle), each acute angle is equal to 45°. There are special relationships between the legs and the hypotenuse:

$$\text{each leg} = \tfrac{1}{2}(\text{hypotenuse})\sqrt{2}$$
$$\text{hypotenuse} = (\text{leg})\sqrt{2}$$

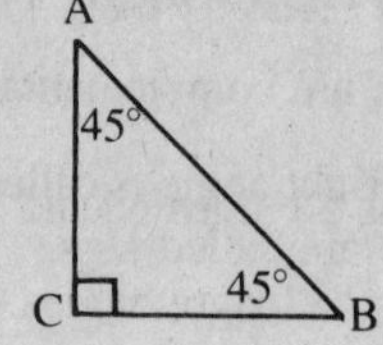

$$AC = BC = \tfrac{1}{2}(AB)\sqrt{2}$$
$$AB = (AC)\sqrt{2} = (BC)\sqrt{2}$$

Example: In isosceles right triangle RST,

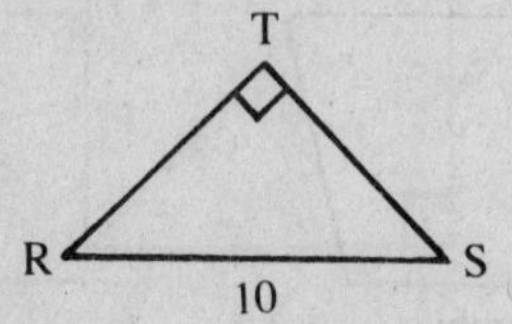

$$RT = \tfrac{1}{2}(10)\sqrt{2}$$
$$= 5\sqrt{2}$$
$$ST = RT = 5\sqrt{2}$$

b In a right triangle with acute angles of 30° and 60°, the leg opposite the 30° angle is one-half the hypotenuse. The leg opposite the 60° angle is one-half the hypotenuse multiplied by $\sqrt{3}$.

Example:

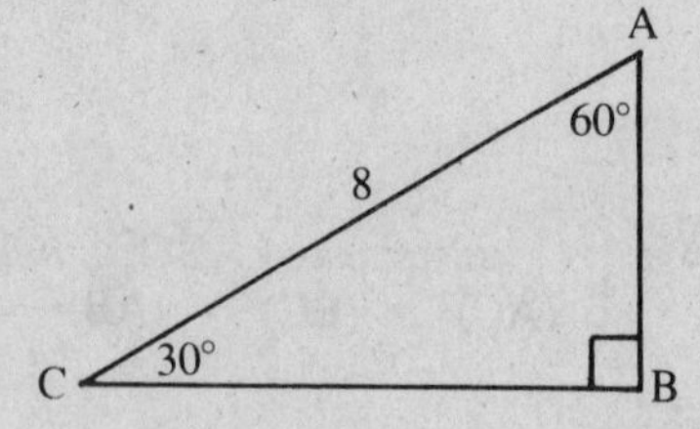

$$AB = \tfrac{1}{2}(8) = 4$$
$$BC = \tfrac{1}{2}(8)\sqrt{3} = 4\sqrt{3}$$

Quadrilaterals

15. a. A **quadrilateral** is a closed, four-sided figure in two dimensions. Common quadrilaterals are the **parallelogram**, **rectangle**, and **square**.

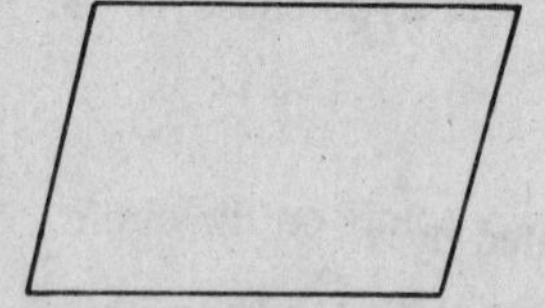

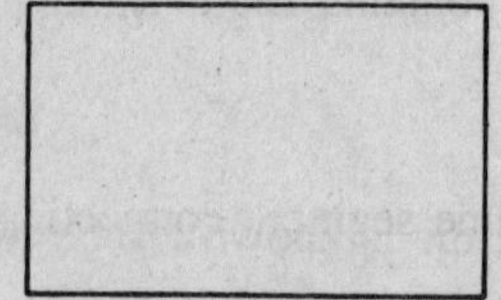

b. The sum of the four angles of a quadrilateral is 360°.

16. a. A **parallelogram** is a quadrilateral in which both pairs of opposite sides are parallel.

b. Opposite sides of a parallelogram are equal.

c. Opposite angles of a parallelogram are equal.

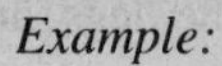

Example:

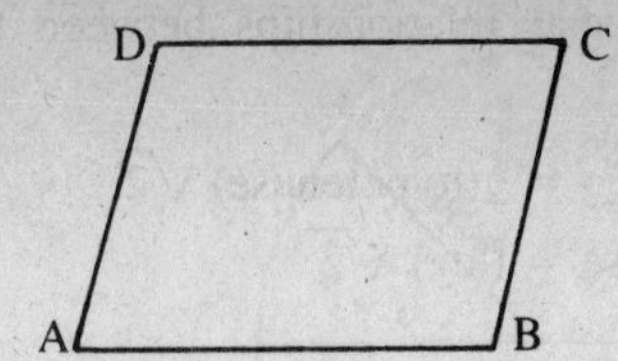

In parallelogram ABCD,
AB || CD, AB = CD, $\angle A = \angle C$
AD || BC, AD = BC, $\angle B = \angle D$

17. a. A **rhombus** is a parallelogram that has all sides equal.

    b. A **rectangle** is a parallelogram that has all right angles.

    c. A **square** is a rectangle that has all sides equal. A square is also a rhombus.

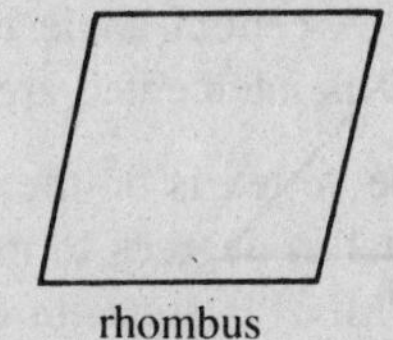

rhombus

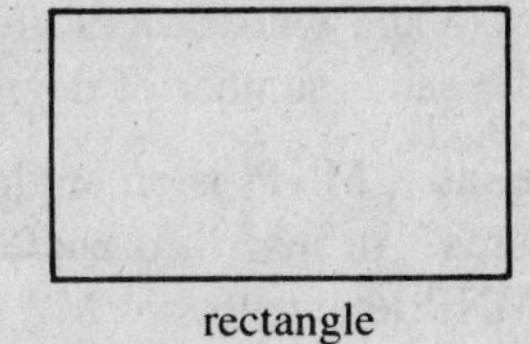

rectangle

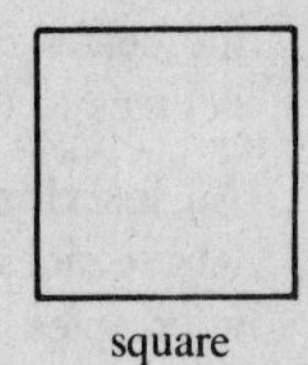

square

18. A **trapezoid** is a quadrilateral with one and only one pair of opposite sides parallel.

In trapezoid ABCD, AB || CD

Circles

19. A **circle** is a closed plane curve, all points of which are equidistant from a point within called the center.

20. a. A **complete circle** contains 360°.

    b. A **semi-circle** contains 180°.

21. a. A **chord** is a line segment connecting any two points on the circle.

    b. A **radius** of a circle is a line segment connecting the center with any point on the circle.

    c. A **diameter** is a chord passing through the center of the circle.

    d. A **secant** is a chord extended in either one or both directions.

    e. A **tangent** is a line touching a circle at one point and only one.

    f. The **circumference** is the curved line bounding the circle.

g. An **arc** of a circle is any part of the circumference.

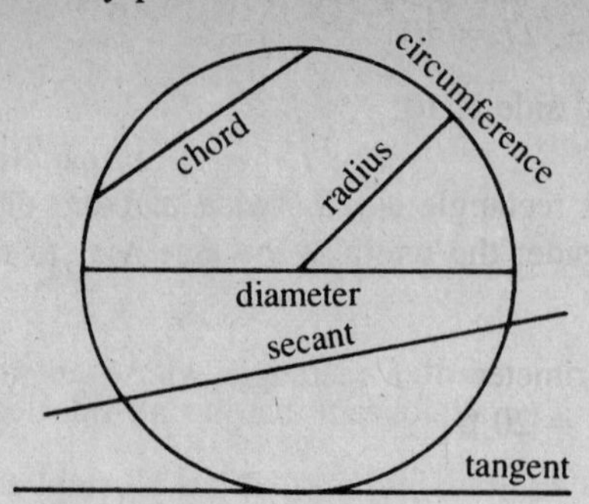

22. a. A **central angle**, as ∡AOB in the figure below, is an angle whose vertex is the center of the circle and whose sides are radii. A central angle is equal in degrees to (or has the same number of degrees as) its intercepted arc.

    b. An **inscribed angle**, as ∡MNP, is an angle whose vertex is on the circle and whose sides are chords. An inscribed angle is equal in degrees to one-half its intercepted arc. ∡MNP intercepts arc MP and ∡MNP is equal in degrees to one-half arc MP.

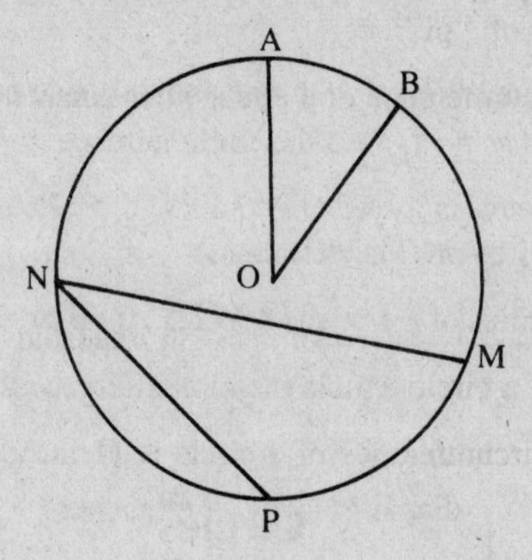

Perimeter

23. The **perimeter** of a two-dimensional figure is the distance around the figure.

    Example: The perimeter of the figure below is $9 + 8 + 4 + 3 + 5 = 29$

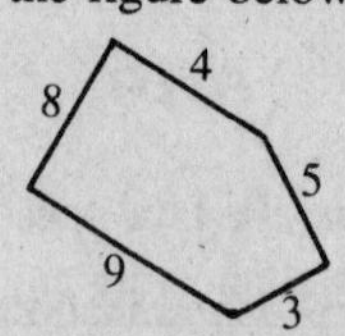

24. a. The perimeter of a triangle is found by adding all of its sides.

    Example: If the sides of a triangle are 4, 5 and 7, its perimeter is $4 + 5 + 7 = 16$.

    b. If the perimeter and two sides of a triangle are given, the third side is found by adding the two given sides and subtracting this sum from the perimeter.

    Illustration: Two sides of a triangle are 12 and 15. The perimeter is 37. Find the other side.

SOLUTION: $12 + 15 = 27$
$37 - 27 = 10$

Answer: The third side is 10.

25. The perimeter of a rectangle equals twice the sum of the length and the width. The length is any side; the width is the side next to the length. The formula is: $P = 2(l + w)$.

Example: The perimeter of a rectangle whose length is 7 feet and width is 3 feet equals $2 \times 10 = 20$ ft.

26. The perimeter of a square equals one side multiplied by 4. The formula is: $P = 4s$.

Example: The perimeter of a square one side of which is 5 feet equals 4×5 feet $= 20$ feet.

27. a. The **circumference** of a circle is equal to the product of the diameter multiplied by π. The formula is $C = \pi d$.

b. The number π ("pi") is approximately equal to $\frac{22}{7}$, or 3.14 (3.1416 for greater accuracy). The problem will state which value to use; otherwise, express the answer in terms of "pi," π.

Example: The circumference of a circle whose diameter is 4 inches $= 4\pi$ inches; or, if it is stated that $\pi = \frac{22}{7}$, then the circumference $= 4 \times \frac{22}{7} = \frac{88}{7} = 12\frac{4}{7}$ inches.

c. Since the diameter is twice the radius, the circumference equals twice the radius multiplied by π. The formula is $C = 2\pi r$.

Example: If the radius of a circle is 3 inches, then the circumference $= 6\pi$ inches.

d. The diameter of a circle equals the circumference divided by π.

Example: If the circumference of a circle is 11 inches, then, assuming $\pi = \frac{22}{7}$,

$$\text{diameter} = 11 \div \frac{22}{7} \text{ inches}$$
$$= \overset{1}{\cancel{11}} \times \frac{7}{\underset{2}{\cancel{22}}} \text{ inches}$$
$$= \frac{7}{2} \text{ inches, or } 3\frac{1}{2} \text{ inches}$$

Area

28. a. In a figure of two dimensions, the total space within the figure is called the **area**.

b. Area is expressed in square denominations, such as **square inches, square centimeters** and **square miles**.

c. In computing area, all dimensions must be in the same denomination.

29. The area of a square is equal to the square of the length of any side. The formula is $A = s^2$.

Example: The area of a square one side of which is 6 inches is $6 \times 6 = 36$ square inches.

30. a. The area of a rectangle equals the product of the length multiplied by the width. The formula is $A = l \times w$.

Example: If the length of a rectangle is 6 feet and its width is 4 feet, then the area is $6 \times 4 = 24$ square feet.

b. If given the area of a rectangle and one dimension, you can find the other dimension by dividing the area by the given dimension.

Example: If the area of a rectangle is 48 square feet and one dimension is 4 feet, then the other dimension is $48 \div 4 = 12$ feet.

31. a. The altitude, or **height**, of a parallelogram is a line drawn from a vertex perpendicular to the opposite side, or **base**.

Example:

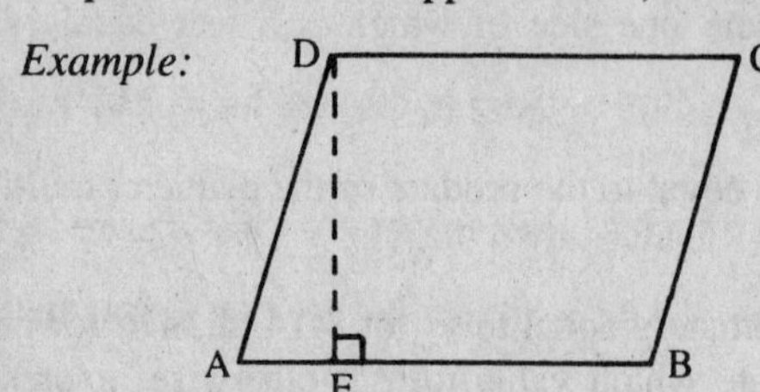

DE is the height
AB is the base

b. The area of a parallelogram is equal to the product of its base and its height. The formula is $A = b \times h$.

Example: If the base of a parallelogram is 10 centimeters and its height is 5 centimeters, its area is $10 \times 5 = 50$ square centimeters.

c. To find the base or the height of a parallelogram given one of these dimensions and given the area, divide the area by the given dimension.

Example: If the area of a parallelogram is 40 square inches and its height is 8 inches, its base is $40 \div 8 = 5$ inches.

32. a. The altitude, or height, of a triangle is a line drawn from a vertex perpendicular to the opposite side, called the base.

b. The area of a triangle is equal to one-half the product of the base and the height. The formula is $A = \frac{1}{2}b \times h$.

Example: The area of a triangle that has a height of 5 inches and a base of 4 inches is $\frac{1}{2} \times 5 \times 4 = \frac{1}{2} \times 20 = 10$ square inches.

c. In a right triangle, one leg may be considered the height and the other leg the base. Therefore, the area of a right triangle is equal to one-half the product of the legs.

Example: The legs of a right triangle are 3 and 4. Its area is $\frac{1}{2} \times 3 \times 4 = 6$ square units.

33. The area of a rhombus is equal to one-half the product of its diagonals. The formula is: $A = \frac{1}{2} \cdot d_1 \cdot d_2$.

Example: If the diagonals of a rhombus are 4 and 6,

$$\text{Area} = \tfrac{1}{2} \cdot 4 \cdot 6$$
$$= 12$$

34. The area of a trapezoid is equal to one-half the product of the height and the sum of the bases.

$$\text{Area} = \tfrac{1}{2}h(\text{base}_1 + \text{base}_2)$$

Example: The area of trapezoid ABCD $= \frac{1}{2} \cdot 4 \cdot (5 + 10)$
$= 2 \cdot 15$
$= 30$

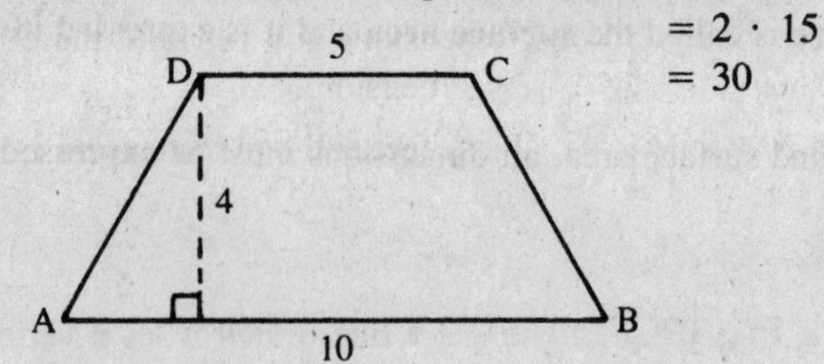

35. a. The area of a circle is equal to the radius squared multiplied by π. The formula is $A = \pi r^2$.

Example: If the radius of a circle is 6 inches, then the area $= 36\pi$ square inches.

b. To find the radius of a circle given the area, divide the area by π and find the square root of the quotient.

Example: To find the radius of a circle of area 100π:

$$\frac{100\pi}{\pi} = 100$$

$$\sqrt{100} = 10 = \text{radius}$$

36. Some figures are composed of several geometric shapes. To find the area of such figures, it is necessary to find the area of each of their parts.

Illustration: Find the area of the figure below:

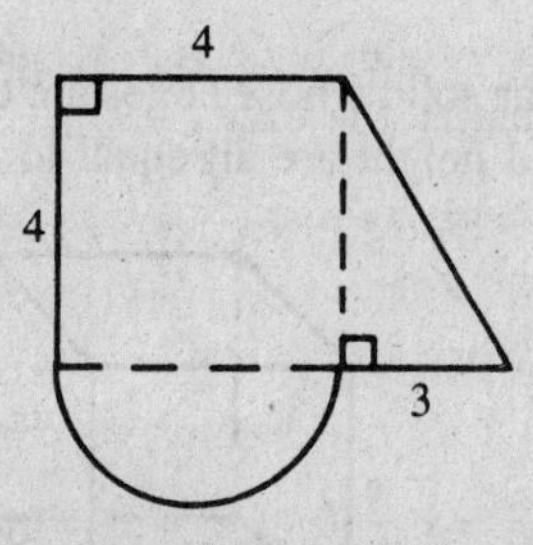

SOLUTION: The figure is composed of three parts: a square of side 4, a semi-circle of diameter 4 (the lower side of the square), and a right triangle with legs 3 and 4 (the right side of the square).

$$\text{Area of square} = 4^2 = 16$$

$$\text{Area of triangle} = \tfrac{1}{2} \times 3 \times 4 = 6$$

$$\text{Area of semi-circle is } \tfrac{1}{2} \text{ area of circle} = \tfrac{1}{2}\pi r^2$$

$$\text{Radius} = \tfrac{1}{2} \times 4 = 2$$

$$\text{Area} = \tfrac{1}{2}\pi r^2$$

$$= \tfrac{1}{2}\pi 2^2 = 2\pi$$

$$\text{Total area} = 16 + 6 + 2\pi = 22 + 2\pi$$

Three-Dimensional Figures

37. a. In a three dimensional figure, the total space contained within the figure is called the **volume** and is expressed in cubic denominations.

    b. The total outside surface is called the **surface area** and it is expressed in square denominations.

    c. In computing volume and surface area, all dimensions must be expressed in the same denomination.

38. a. A rectangular solid is a figure of three dimensions having six rectangular faces meeting each other at right angles. The three dimensions are **length**, **width**, and **height**. The figure below is a rectangular solid; "l" is the length, "w" is the width, and "h" is the height.

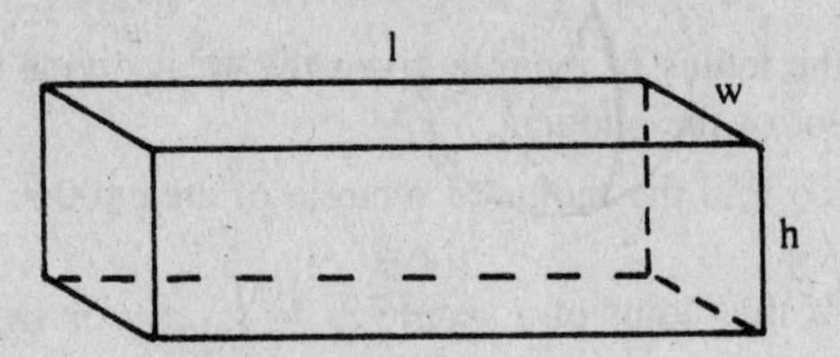

    b. The volume of a rectangular solid is the product of the length, width, and height; $V = l \times w \times h$.

    Example: The volume of a rectangular solid whose length is 6 ft, width 3 ft, and height 4 ft is $6 \times 3 \times 4 = 72$ cubic ft.

39. a. A **cube** is a rectangular solid whose edges are equal. The figure below is a cube; the length, width, and height are all equal to "e."

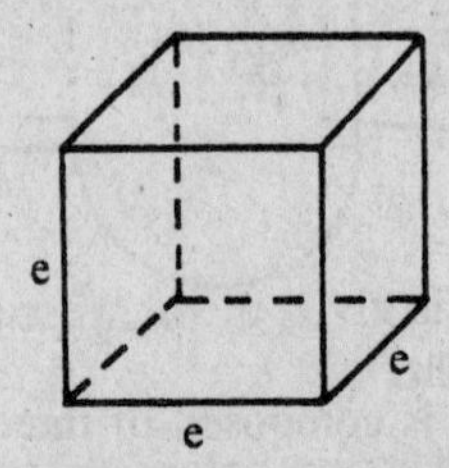

    b. The volume of a cube is equal to the edge cubed; $v = e^3$.

    Example: The volume of a cube whose height is 6 inches equals $6^3 = 6 \times 6 \times 6 = 216$ cubic inches.

    c. The surface area of a cube is equal to the area of any side multiplied by 6.

    Example: The surface area of a cube whose length is 5 inches $= 5^2 \times 6 = 25 \times 6 = 150$ square inches.

40. The volume of a circular cylinder is equal to the product of π, the radius squared, and the height.

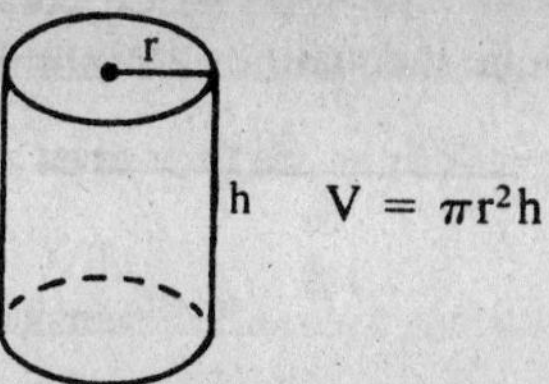

Example: A circular cylinder has a radius of 7 inches and a height of $\frac{1}{2}$ inch. Using $\pi = \frac{22}{7}$, its volume is:

$$\frac{22}{7} \times 7 \times 7 \times \frac{1}{2} = 77 \text{ cubic inches}$$

41. The volume of a sphere is equal to $\frac{4}{3}$ the product of π and the radius cubed.

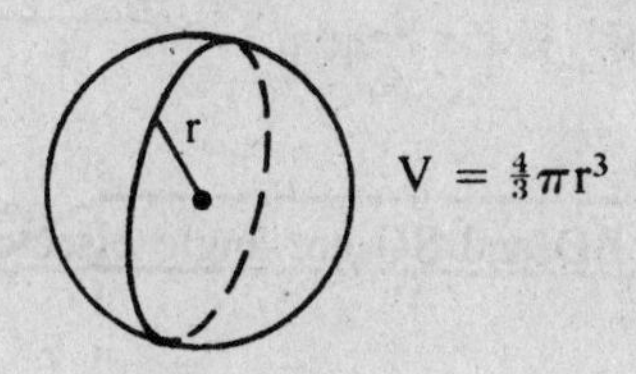

Example: If the radius of a sphere is 3 cm, its volume in terms of π is:

$$\frac{4}{3} \times \pi \times 3 \times 3 \times 3 = 36\pi \text{ cubic centimeters}$$

42. The volume of a cone is given by the formula $V = \frac{1}{3}\pi r^2 h$, where r is the radius and h is the height.

Example: In the cone shown below, if $h = 9$ cm, $r = 10$ cm and $\pi = 3.14$, then the volume is:

$$\frac{1}{3} \times 3.14 \times 10 \times 10 \times 9 \text{ cm}^3 = 3.14 \times 300 \text{ cm}^3$$
$$= 942 \text{ cm}^3$$

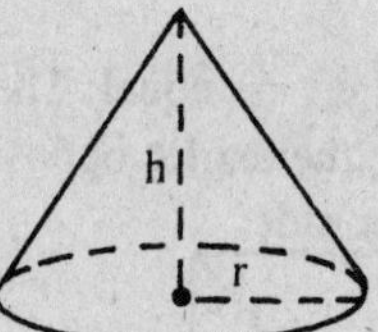

43. The volume of a pyramid is given by the formula $V = \frac{1}{3}Bh$, where B is the area of the base and h is the height.

Example: In the pyramid shown below, the height is 10 inches and the side of the base is 3 inches. Since the base is a square,

$$B = 3^2 = 9 \text{ square inches}$$
$$V = \frac{1}{3} \times 9 \times 10 = 30 \text{ cubic inches}$$

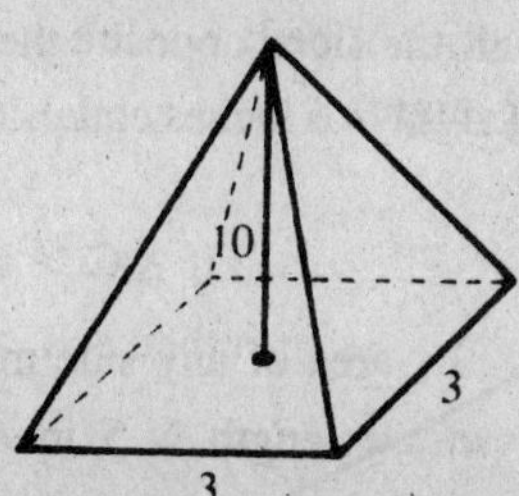

Geometric Inequalities

44. The basic principles of algebraic inequalities are true for geometric figures.

Example:

A B C D

If AB < CD, then AB + BC < BC + CD, or AC < BD

Example:

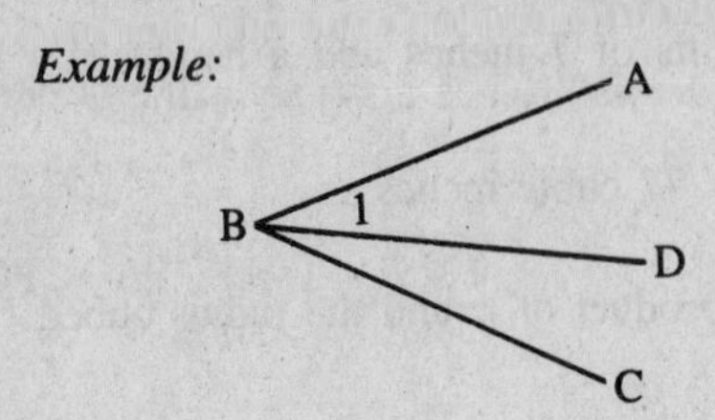

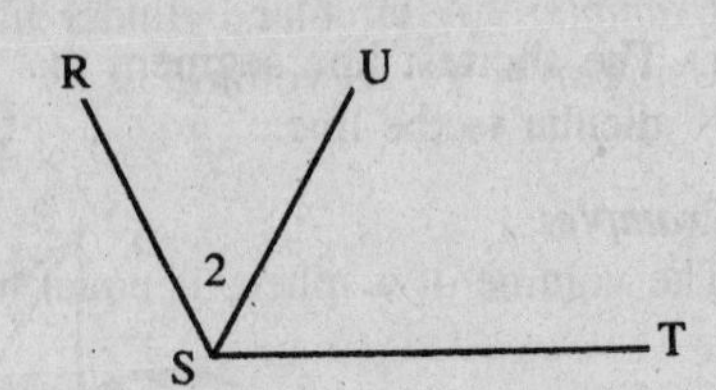

If $\angle$RST > $\angle$ABC, and BD and SU are angle bisectors (divide the angles in half), then

$$\tfrac{1}{2}\angle RST > \tfrac{1}{2}\angle ABC, \text{ or } \angle 2 > \angle 1$$

45. a. In **triangle inequalities**, the sum of two sides of a triangle is greater than the third side.

Example:

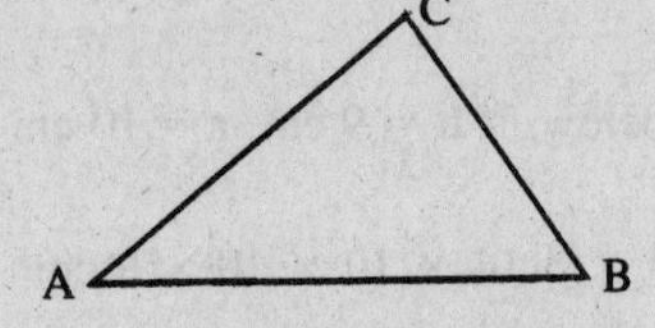

AC + CB > AB

b. If two sides of a triangle are unequal, the angles opposite them are unequal, and the larger angle is opposite the larger side.

Example:

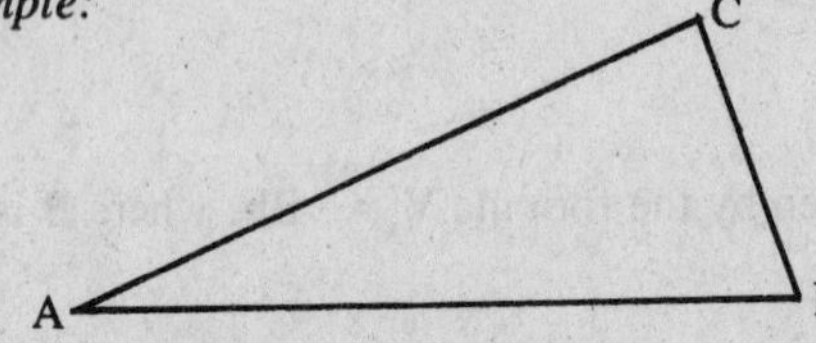

If AC > CB,
then $\angle$B > $\angle$A

c. If two angles of a triangle are unequal, the sides opposite them are unequal, and the larger side is opposite the larger angle.

Example:

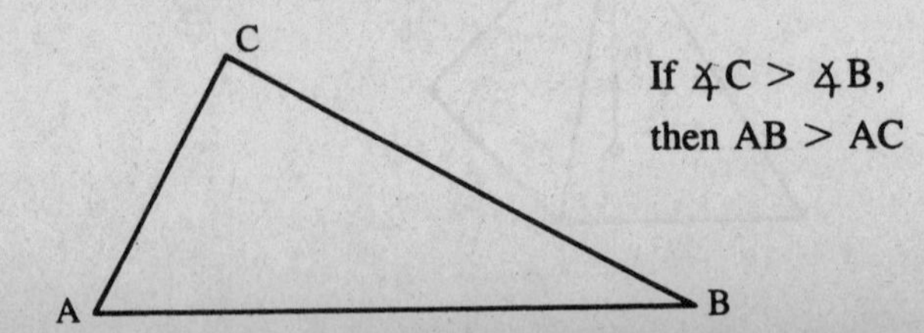

If $\angle$C > $\angle$B,
then AB > AC

d. An exterior angle of a triangle is greater than either remote interior angle.

Example:

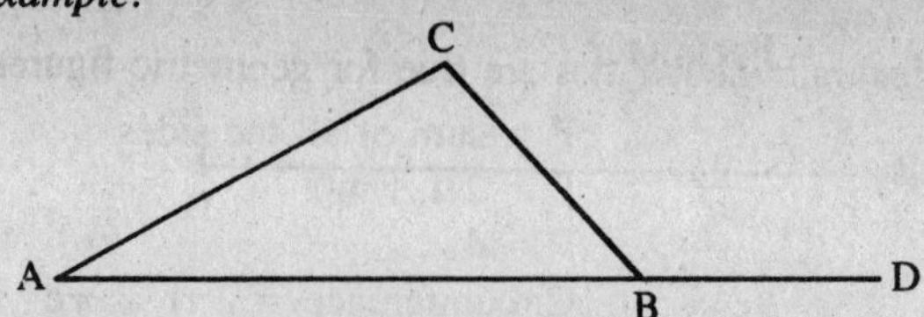

∡CBD is an exterior angle
∡CBD > ∡A
∡CBD > ∡C

e. The shortest line segment that can be drawn from a point to a line is perpendicular to the line.

Example:

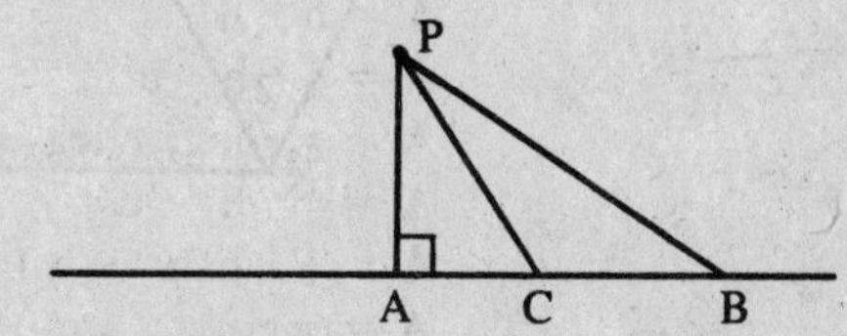

If PA ⊥ AB, PA < PC
and PA < PB

Summary of Geometric Formulas

Perimeter

Any 2-dimensional figure	P = sum of all the sides
Rectangle	$P = 2(l + w)$
Square	$P = 4s$
Circle	Circumference $= 2\pi r = \pi d$

Area

Square	$A = s^2$
Rectangle	$A = l \cdot w$
Parallelogram	$A = b \cdot h$
Triangle	$A = \frac{1}{2} \cdot b \cdot h$
Right triangle	$A = \frac{1}{2} \cdot leg_1 \cdot leg_2$
Rhombus	$A = \frac{1}{2} \cdot d_1 \cdot d_2$
Trapezoid	$A = \frac{1}{2} \cdot h(b_1 + b_2)$
Circle	$A = \pi r^2$

Volume

Rectangular solid	$V = l \cdot w \cdot h$
Cube	$V = c^3$
Circular cylinder	$V = \pi r^2 h$
Sphere	$V = \frac{4}{3}\pi r^3$
Cone	$V = \frac{1}{3}\pi r^2 h$
Pyramid	$V = \frac{1}{3} \cdot B \cdot h$ (B = area of base)

Practice Problems Involving Geometry

1. If the perimeter of a rectangle is 68 yards and the width is 48 feet, the length is
   (A) 10 yards (C) 20 feet
   (B) 18 yards (D) 56 feet

2. The total length of fencing needed to enclose a rectangular area 46 feet by 34 feet is
   (A) 26 yards 1 foot (C) 52 yards 2 feet
   (B) $26\frac{2}{3}$ yards (D) $53\frac{1}{3}$ yards

3. An umbrella 50″ long can lie on the bottom of a trunk whose length and width are, respectively
   (A) 36″, 30″ (C) 42″, 36″
   (B) 42″, 24″ (D) 39″, 30″

4. A road runs 1200 ft. from A to B, and then makes a right angle going to C, a distance of 500 ft. A new road is being built directly from A to C. How much shorter will the new road be?
   (A) 400 ft. (C) 850 ft.
   (B) 609 ft. (D) 1300 ft.

5. A certain triangle has sides that are, respectively, 6 inches, 8 inches, and 10 inches long. A rectangle equal in area to that of the triangle has a width of 3 inches. The perimeter of the rectangle, expressed in inches, is
   (A) 11 (C) 22
   (B) 16 (D) 24

6. If AB || DE, $\measuredangle C = 50°$ and $\measuredangle 1 = 60°$, then $\measuredangle A =$
   (A) 30°
   (B) 60°
   (C) 70°
   (D) 50°

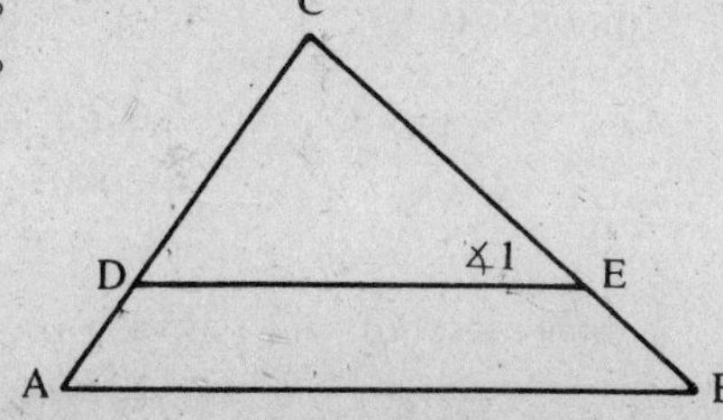

7. A rectangular bin 4 feet long, 3 feet wide, and 2 feet high is solidly packed with bricks whose dimensions are 8 inches, 4 inches, and 2 inches. The number of bricks in the bin is
   (A) 54 (C) 1296
   (B) 648 (D) none of these

8. If the cost of digging a trench is \$2.12 a cubic yard, what would be the cost of digging a trench 2 yards by 5 yards by 4 yards?
   (A) \$21.20 (C) \$64.00
   (B) \$40.00 (D) \$84.80

9. A piece of wire is shaped to enclose a square, whose area is 121 square inches. It is then reshaped to enclose a rectangle whose length is 13 inches. The area of the rectangle, in square inches, is
   (A) 64 (C) 117
   (B) 96 (D) 144

10. The area of a 2-foot-wide walk around a garden that is 30 feet long and 20 feet wide is
    (A) 104 sq. ft. (C) 680 sq. ft.
    (B) 216 sq. ft. (D) 704 sq. ft.

11. The area of a circle is 49π. Find its circumference, in terms of π.
    (A) 14π (C) 49π
    (B) 28π (D) 98π

12. In two hours, the minute hand of a clock rotates through an angle of
    (A) 90° (C) 360°
    (B) 180° (D) 720°

13. A box is 12 inches in width, 16 inches in length, and 6 inches in height. How many square inches of paper would be required to cover it on all sides?
    (A) 192 (C) 720
    (B) 360 (D) 1440

14. If the volume of a cube is 64 cubic inches, the sum of its edges is

(A) 48 inches
(B) 32 inches
(C) 16 inches
(D) 24 inches

15. The diameter of a conical pile of cement is 30 feet and its height is 14 feet. If $\frac{3}{4}$ cubic yard of cement weighs 1 ton, the number of tons of cement in the cone to the nearest ton is

(Volume of a cone $= \frac{1}{3}\pi r^2 h$; use $\pi = \frac{22}{7}$)

(A) 92
(B) 163
(C) 489
(D) 652

16. In triangle ABC, if $\measuredangle A < \measuredangle B$ and $\measuredangle B < \measuredangle C$, then

(A) $\measuredangle A > \measuredangle C$
(B) $\measuredangle A = \measuredangle C$
(C) AB < BC
(D) AB > AC

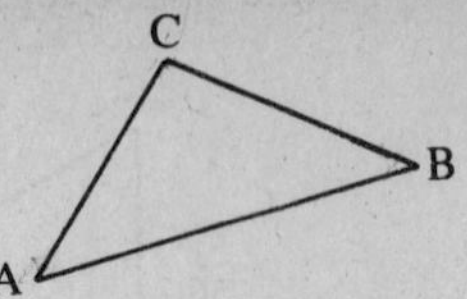

17. If AC ⊥ BD, which of the following must be true?

I. BC < BD
II. BC < AB
III. AD = DC
IV. $\measuredangle A = \measuredangle C$

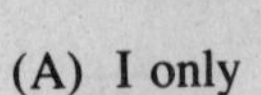

(A) I only
(B) II only
(C) III and IV only
(D) I, II, III and IV

Geometry Problems — Correct Answers

1. (B)
2. (D)
3. (C)
4. (A)
5. (C)
6. (C)
7. (B)
8. (D)
9. (C)
10. (B)
11. (A)
12. (D)
13. (C)
14. (A)
15. (B)
16. (D)
17. (B)

Problem Solutions — Geometry

1. Perimeter = 2(l + w). Let the length be x yards.

$$\text{Each width} = 48 \text{ ft} = 16 \text{ yd}$$

$$2(x + 16) = 68$$
$$2x + 32 = 68$$
$$-32 \quad -32$$
$$\frac{2x}{2} = \frac{36}{2}$$
$$x = 18$$

Answer: **(B)** 18 yards

2. Perimeter $= 2(l + w)$
$= 2(46 + 34)$ feet
$= 2 \times 80$ feet
$= 160$ feet

160 feet $= 160 \div 3$ yards $= 53\frac{1}{3}$ yards

Answer: **(D)** $53\frac{1}{3}$ yards

3. The umbrella would be the hypotenuse of a right triangle whose legs are the dimensions of the trunk. According to the Pythagorean Theorem, in any right triangle the square of the hypotenuse equals the sum of the squares of the legs. Therefore, the sum of the dimensions squared must at least equal the length of the umbrella squared: $(50)^2 = 2500$.

The only set of dimensions that fills this condition is (C):

$$(42)^2 + (36)^2 = 1764 + 1296 = 3060$$

Answer: **(C)** 42″, 36″

4. The new road is the hypotenuse of a right triangle whose legs are the old road.

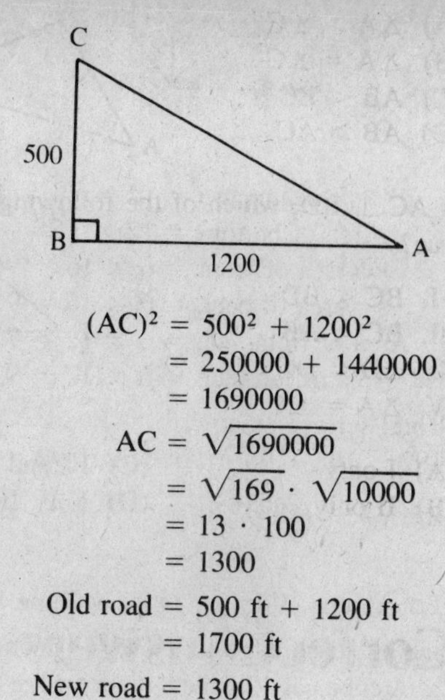

$$(AC)^2 = 500^2 + 1200^2$$
$$= 250000 + 1440000$$
$$= 1690000$$
$$AC = \sqrt{1690000}$$
$$= \sqrt{169} \cdot \sqrt{10000}$$
$$= 13 \cdot 100$$
$$= 1300$$
$$\text{Old road} = 500 \text{ ft} + 1200 \text{ ft}$$
$$= 1700 \text{ ft}$$
$$\text{New road} = 1300 \text{ ft}$$
$$\text{Difference} = 400 \text{ ft}$$

Answer: **(A)** 400 feet

5. Since $6^2 + 8^2 = 10^2$, or $36 + 64 = 100$, the triangle is a right triangle. Its area is $\frac{1}{2} \times 6 \times 8 = 24$ sq. in. (area of a triangle $= \frac{1}{2} \cdot b \cdot h$). Therefore, the area of the rectangle is also 24 square inches. If the width of the rectangle is 3 inches, the length is $24 \div 3 = 8$ inches. Then, the perimeter of the rectangle is $2(3 + 8) = 2 \times 11 = 22$ inches.

Answer: **(C)** 22

6. $\angle B$ and $\angle 1$ are corresponding angles formed by the parallel lines AB and DE and the transversal BC. Therefore, $\angle 1 = \angle B = 60°$.

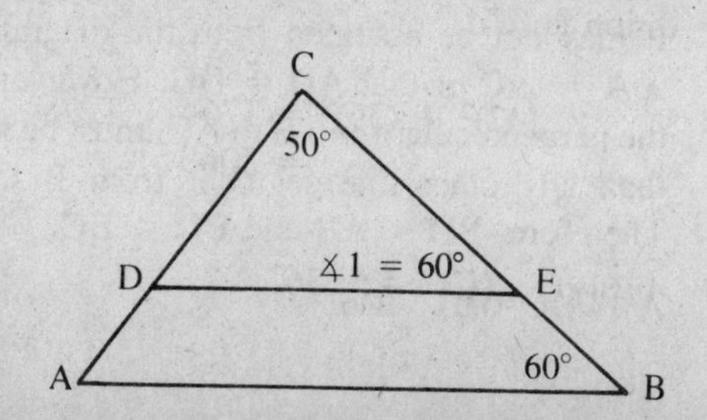

The sum of the angles of a triangle is 180°.

$$\angle A + \angle B + \angle C = 180°$$
$$\angle A + 60° + 50° = 180°$$
$$\angle A + 110° = 180°$$
$$-110° \quad -110°$$
$$\angle A = 70°$$

Answer: **(C)** 70°

7. Convert the dimensions of the bin to inches:

4 feet = 48 inches
3 feet = 36 inches
2 feet = 24 inches

Volume of bin $= 48 \times 36 \times 24$ cubic inches
$= 41{,}472$ cubic inches

Volume of each brick $= 8 \times 4 \times 2$ cubic inches
$= 64$ cubic inches

$41472 \div 64 = 648$ bricks

Answer: **(B)** 648

8. The trench contains:

$2 \text{ yd} \times 5 \text{ yd} \times 4 \text{ yd} = 40$ cubic yards
$40 \times \$2.12 = \$84.80$

Answer: **(D)** \$84.80

9. If the area of the square is 121 square inches, each side is $\sqrt{121} = 11$ inches and the perimeter is $4 \times 11 = 44$ inches. The perimeter of the rectangle is then 44 inches. If the two lengths are each 13 inches, their total is 26 inches. $44 - 26 = 18$ inches remain for the two widths. Therefore, each width is equal to $18 \div 2 = 9$ inches.

The area of a rectangle with length 13 inches and width 9 inches is $13 \times 9 = 117$ square inches.

Answer: **(C)** 117

10. The walk consists of:

a) 2 rectangles of length 30 ft and width 2 ft.
   Area of each = $2 \times 30 = 60$ sq ft
   Area of both = 120 sq ft

b) 2 rectangles of length 20 ft and width 2 ft.
   Area of each = $2 \times 20 = 40$ sq ft
   Area of both = 80 sq ft

c) 4 squares, each having a side of 2 ft.
   Area of each square = $2^2 = 4$ sq ft
   Area of 4 squares = 16 sq ft

Total area of walk = $120 + 80 + 16$
= 216 sq ft

Alternatively, you may solve this problem by finding the area of the garden and the area of the garden plus the walk, then subtracting to find the area of the walk alone:

Area of garden = $20 \times 30 = 600$ sq ft

Area of garden + walk:
$(20 + 2 + 2) \times (30 + 2 + 2) = 24 \times 34$
= 816 sq ft

Area of walk alone:
$816 - 600 = 216$ sq ft

Answer: **(B)** 216 sq ft

11. Area of a circle = πr^2. If the area is 49π, the radius is $\sqrt{49} = 7$.

Circumference = $2\pi r$
$= 2 \times \pi \times 7$
$= 14\pi$

Answer: **(A)** 14π

12. In one hour, the minute hand rotates through 360°. In two hours it rotates through $2 \times 360° = 720°$.

Answer: **(D)** 720°

13.

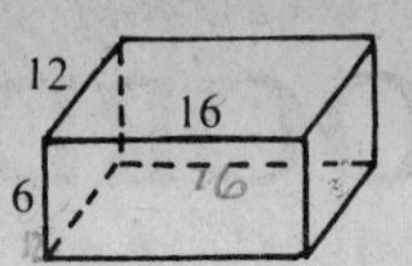

Area of top = $12 \times 16 = 192$ sq in.
Area of bottom = $12 \times 16 = 192$ sq in.
Area of front = $6 \times 16 = 96$ sq in.
Area of back = $6 \times 16 = 96$ sq in.
Area of right side = $6 \times 12 = 72$ sq in.
Area of left side = $6 \times 12 = 72$ sq in.

Total surface area:
$192 + 192 + 96 + 96 + 72 + 72 = 720$ sq in.

Answer: **(C)** 720

14. For a cube, $V = e^3$. If the volume is 64 cubic inches, each edge is $\sqrt[3]{64} = 4$ inches.

A cube has 12 edges. If each edge is 4 inches, the sum of the edges is $4 \times 12 = 48$ inches.

Answer: **(A)** 48 inches

15. If diameter = 30, radius = 15.

$V = \frac{1}{3} \times \frac{22}{7} \times 15 \times 15 \times 14$
= 3300 cubic feet

27 cubic feet = 1 cubic yard
3300 cu ft ÷ 27 cu ft = $122\frac{2}{9}$ cu yd

$122\frac{2}{9} \div \frac{3}{4} = \frac{1100}{9} \times \frac{4}{3} = \frac{4400}{27}$
= 163 tons to the nearest ton

Answer: **(B)** 163

16. ∡C is the largest angle. AB, the side opposite ∡C, must be the largest side of the triangle. Therefore, AB > AC.

Answer: **(D)** AB > AC

17. It may not be assumed from the diagram that ∡A = ∡C or that AD = DC. However, BD, the perpendicular from B to AC, must be shorter than any other line segment from B to AC. Therefore, BD < AB and BD < BC.

Answer: **(B)** II only

COORDINATE GEOMETRY

1. a. **Coordinate geometry** is used to locate and to graph points and lines on a plane.
   b. The coordinate system is made up of two number lines that are perpendicular and that intersect at 0.

   The horizontal number line is called the **x-axis**.

   The vertical number line is called the **y-axis**.

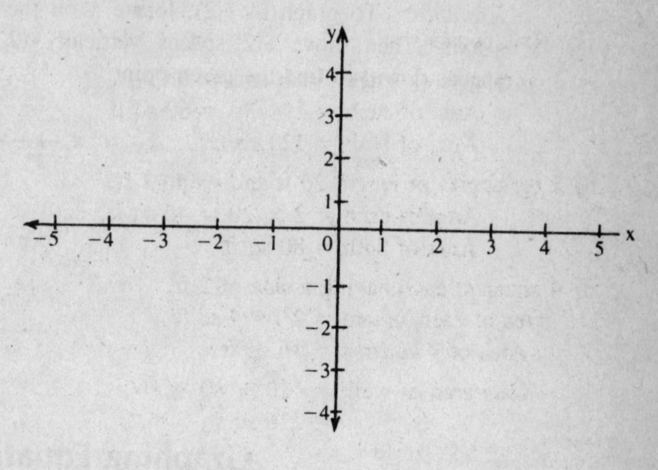

Graphing Points

2. a. Any point on the plane has two numbers **(coordinates)**, which indicate its location. The **x-coordinate (abscissa)** is found by drawing a vertical line from the point to the x-axis. The number on the x-axis where the vertical line meets it is the x-coordinate of the point.

   The **y-coordinate (ordinate)** is found by drawing a horizontal line from the point to the y-axis. The number on the y-axis where the horizontal line meets it is the y-coordinate of the point. The two coordinates are always given in the order (x,y).

Example:

The x-coordinate of point A is 3.

The y-coordinate of point A is 2.

The coordinates of point A are given by the ordered pair (3,2).

Point B has coordinates (−1,4).

Point C has coordinates (−4,−3).

Point D has coordinates (2,−3).

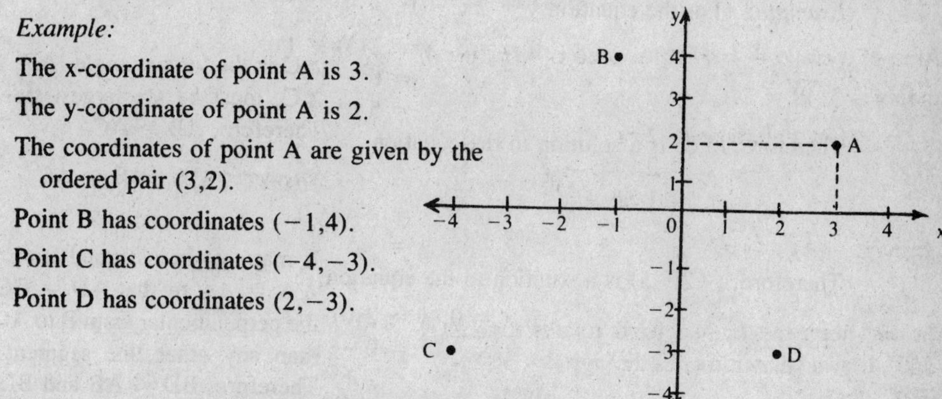

b. The point at which the x-axis and the y-axis meet has coordinates (0,0) and is called the **origin**.

c. Any point on the y-axis has 0 as its x-coordinate. Any point on the x-axis has 0 as its y-coordinate.

3. To graph a point whose coordinates are given, first locate the x-coordinate on the x-axis. From that position, move vertically the number of spaces indicated by the y-coordinate.

Example: To graph (4,−2), locate 4 on the x-axis. Then move −2 spaces vertically (2 spaces down) to find the given point.

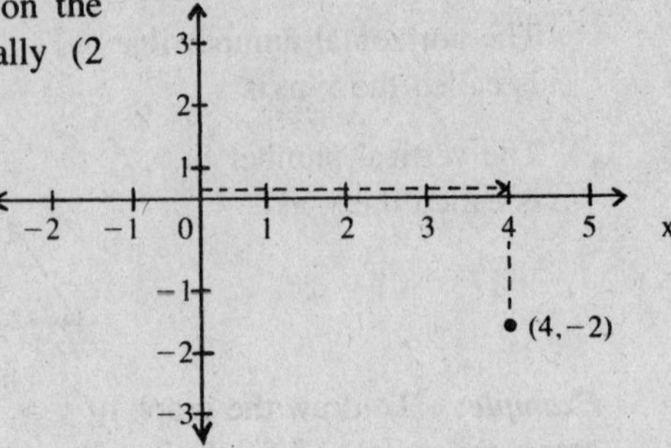

Graphing Equations

4. a. For any equation in x or y or both, ordered pairs (x,y) may be found which are solutions for (or which "satisfy") the equation.

Example: (3,4), (1,6) and (7,0) are solutions to the equation $x + y = 7$, since $3 + 4 = 7$, $1 + 6 = 7$ and $7 + 0 = 7$.

Example: (2,0), (2,1), (2,3) and (2,10) all satisfy the equation $x = 2$. Note that the value of y is irrelevant in this equation.

Example: (−3,1), (4,1) and (12,1) all satisfy the equation $y = 1$.

b. To find ordered pairs that satisfy an equation, it is usually easiest to substitute any value for x and solve the resulting equation for y.

Example: For the equation $y = 2x - 1$:

$$\text{if } x = 3,\ y = 2(3) - 1 = 6 - 1 = 5$$

Therefore, (3,5) is a solution to the equation.

$$\text{if } x = -2,\ y = 2(-2) - 1 = -4 - 1 = -5$$

Therefore, (−2,−5) is a solution to the equation.

$$\text{if } x = 0,\ y = 2(0) - 1 = 0 - 1 = -1$$

Therefore, (0,−1) is a solution to the equation.

c. If two or more ordered pairs that satisfy a given equation are graphed and the points are connected, the resulting line is the graph of the given equation.

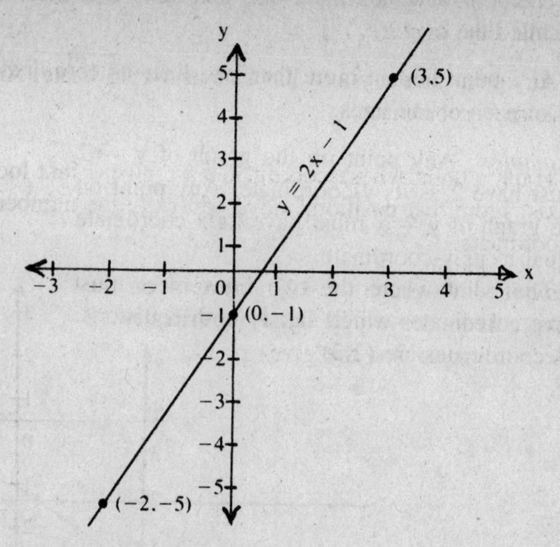

Example: To draw the graph of $y = 2x - 1$, graph the points $(3,5)$, $(-2,-5)$ and $(0,-1)$. Then draw the line passing through all of them.

5. Any equation that can be written in the form $y = mx + b$, where m and b remain constant, is called a **linear equation** and has a straight line as its graph.

   Example: $y = x$ may be written $y = 1x + 0$ and has a straight line graph.

   Example: The equation $y - 3 = 2x$ may be rewritten:

$$\begin{array}{rcl} y - 3 & = & 2x \\ +\,3 & & +\,3 \\ \hline y & = & 2x + 3 \end{array}$$

   Therefore, the graph of $y - 3 = 2x$ is a straight line.

6. a. Any line parallel to the x-axis has the equation $y = a$, where a is constant.

   b. Any line parallel to the y-axis has the equation $x = b$, where b is constant.

   Example: The graph of $y = 5$ is parallel to the x-axis and passes through the y-axis at 5.
   The graph of $x = -1$ is parallel to the y-axis and passes through the x-axis at -1.

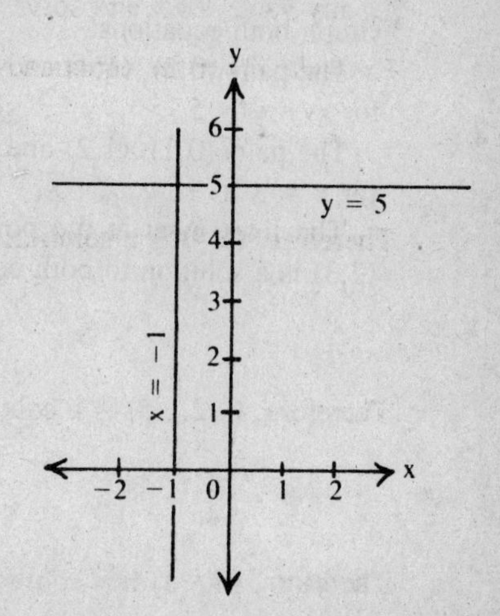

7. a. The coordinates of any point on a straight line must satisfy the equation of that line.

   b. If a point lies on more than one line, its coordinates must satisfy the equation of each of the lines.

Example: Any point on the graph of $y = 2$ must have 2 as its y-coordinate. Any point on the graph of $y = x$ must have its x-coordinate equal to its y-coordinate.

The point where the two lines meet must have coordinates which satisfy both equations. Its coordinates are (2,2).

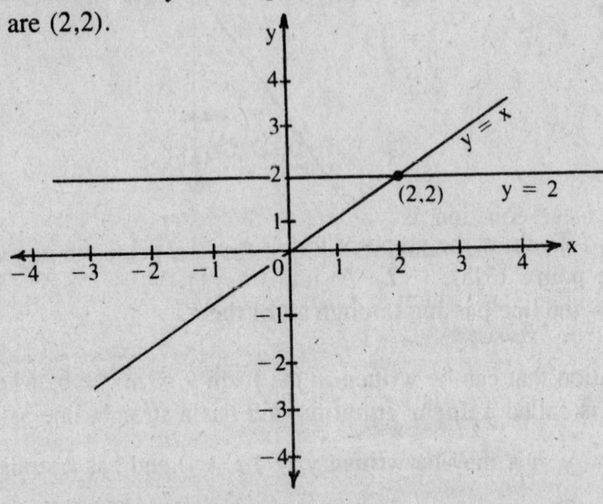

Solving Pairs of Equations

8. To find the ordered pair that is a solution to a pair of equations, graph both equations and find the point at which their corresponding lines meet.

Example: Solve the pair of equations:

$$x + y = 5$$

$$y = x + 1$$

Graph both equations:

The pairs (0,5), (1,4) and (5,0) are solutions for $x + y = 5$.

The pairs (0,1), (1,2) and (3,4) are solutions for $y = x + 1$.

The lines meet at the point (2,3). The pair (2,3) is a solution to both equations.

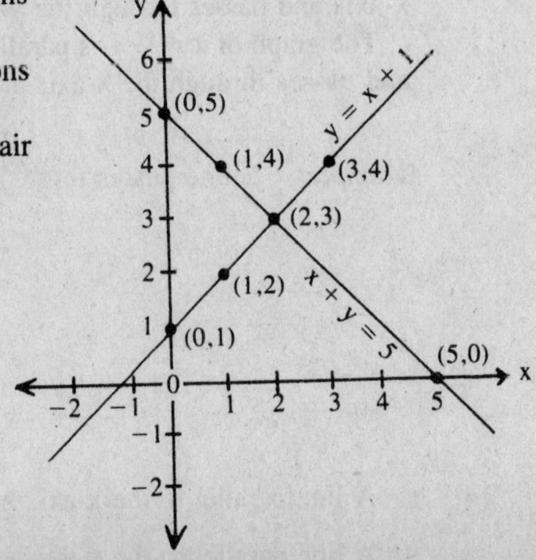

Distance Between Two Points

9. The distance d between any two points (x_1,y_1) and (x_2,y_2) is given by the formula:

$$d = \sqrt{(x_1 - x_2)^2 + (y_1 - y_2)^2}$$

Example: The distance between the points (13,5) and (1,0) is:

$$\begin{aligned} d &= \sqrt{(13 - 1)^2 + (5 - 0)^2} \\ &= \sqrt{(12)^2 + (5)^2} \\ &= \sqrt{144 + 25} \\ &= \sqrt{169} \\ &= 13 \end{aligned}$$

Slope-Intercept

10. When a linear equation is written in the form $y = mx + b$, b is called the **y-intercept** and is the value of y where the line passes through the y-axis.

Example: The line which has the equation $y = 3x - 4$ passes through the y-axis at -4.

11. When a linear equation is written in the form $y = mx + b$, m is called the **slope** of the line.

Example: The line which has the equation $y = 3x - 4$ has slope equal to 3.

12. a. Parallel lines have equal slopes.

Example: $y = -5x - 2$ and $y = -5x + 1$ represent parallel lines which each have slope equal to -5.

b. Perpendicular lines have slopes which are negative reciprocals; that is, their product is -1.

Example: $y = \frac{1}{2}x - 3$ and $y = -2x + 7$ represent perpendicular lines: $(\frac{1}{2})(-2) = -1$.

13. If two points, (x_1,y_1) and (x_2,y_2), are known, the slope m of the line may be found by the formula

$$m = \frac{y_1 - y_2}{x_1 - x_2}$$

Example: A line passes through the points $(4,-5)$ and $(1,-3)$. Its slope is:

$$\begin{aligned} m &= \frac{(-5) - (-3)}{(4) - (1)} \\ &= \frac{-2}{3} \\ &= -\frac{2}{3} \end{aligned}$$

14. a. A line parallel to the x-axis has slope equal to 0.

b. A line parallel to the y-axis is considered to have no slope.

Practice Problems Involving Graphs

1. In the graph below, the coordinates of point A are
(A) $(-1,3)$ (C) $(1,-3)$
(B) $(-3,1)$ (D) $(3,-1)$

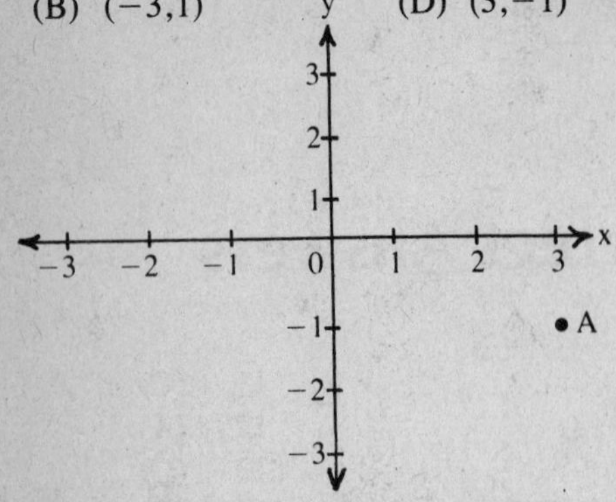

2. A circle has its center at $(0,0)$ and diameter AB. If the coordinates of A are $(-4,0)$, then the coordinates of B are
(A) $(4,0)$ (C) $(0,-4)$
(B) $(0,4)$ (D) $(4,-4)$

3. Point R lies on the graph of $y = 3x - 4$. If the abscissa of R is 1, the ordinate of R is
(A) -1 (C) 3
(B) 1 (D) 4

4. The lines $y = 4$ and $x = 7$ intersect at the point
(A) $(4,7)$ (C) $(3,0)$
(B) $(7,4)$ (D) $(0,3)$

5. The distance from point A to point B in the graph below is
(A) 3 (C) 5
(B) 4 (D) 6

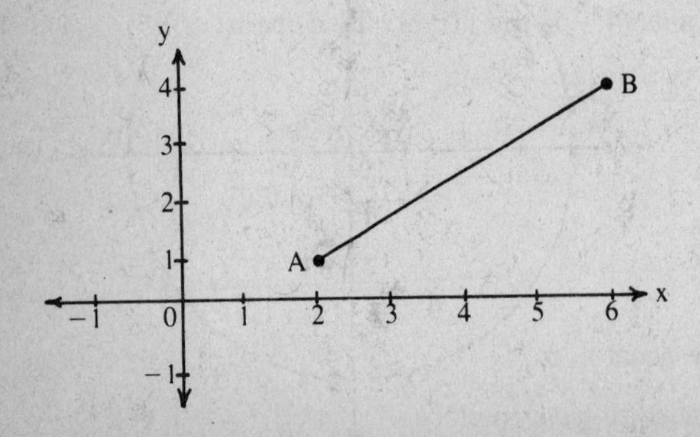

6. The line shown in the graph below has which of the following equations?
(A) $x = 2$ (C) $y = x + 2$
(B) $y = -2$ (D) $x = y + 2$

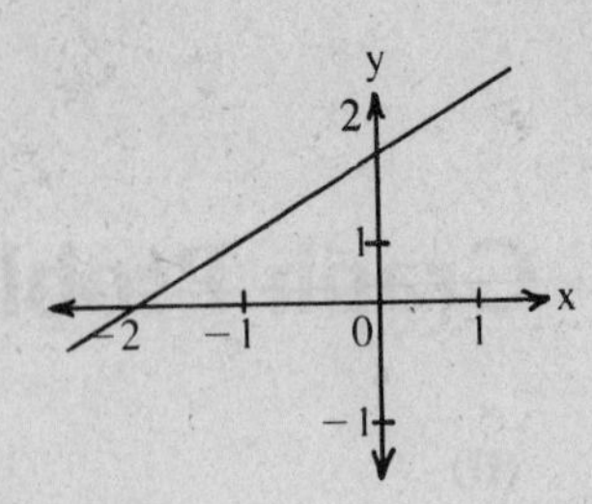

7. The graph of $x + 2y = 6$ intersects the y-axis at the point
(A) $(0,3)$ (C) $(3,0)$
(B) $(0,-3)$ (D) $(-3,0)$

8. The graphs of $y = 2x$ and $y = x + 1$ intersect at the point
(A) $(1,2)$ (C) $(2,1)$
(B) $(0,1)$ (D) $(1,0)$

9. The distance from $(-1,0)$ to $(5,-2)$ is
(A) $4\frac{1}{2}$ (C) $\sqrt{10}$
(B) $5\frac{1}{2}$ (D) $2\sqrt{10}$

10. A triangle has vertices A(1,2), B(11,2) and C(4,5). How many square units are in the area of triangle ABC?
(A) 15 (C) 25
(B) 20 (D) 30

11. The line whose equation is $2x + y = 6$ has slope
(A) 6 (C) -2
(B) 3 (D) 2

12. The y-intercept of the line represented by the equation $2x + y = 6$ is
(A) 6 (C) -2
(B) 3 (D) 2

13. The slope of the line passing through the points $(-3,0)$ and $(3,5)$ is
(A) $\frac{6}{5}$ (C) $\frac{5}{6}$
(B) $-\frac{6}{5}$ (D) 0

Questions 14 and 15 refer to the following equations:

I: $y = 3x - 4$
II: $y = -3x - 4$
III: $y = \frac{1}{3}x - 4$
IV: $y = 3x + 4$

14. Which equations represent parallel lines?
(A) I and II only
(B) II and III only
(C) I and IV only
(D) I, II and IV only

15. Which equations represent perpendicular lines?
(A) I and II
(B) I and IV
(C) II and III
(D) I and III

Graph Problems — Correct Answers

1. **(D)**
2. **(A)**
3. **(A)**
4. **(B)**
5. **(C)**
6. **(C)**
7. **(A)**
8. **(A)**
9. **(D)**
10. **(A)**
11. **(C)**
12. **(A)**
13. **(C)**
14. **(C)**
15. **(C)**

Problem Solutions — Graphs

1. A vertical line through A meets the x-axis at 3; therefore, the x-coordinate is 3.
   A horizontal line through A meets the y-axis at -1; therefore, the y-coordinate is -1.
   The coordinates of point A are $(3,-1)$.

   Answer: **(D)** $(3,-1)$

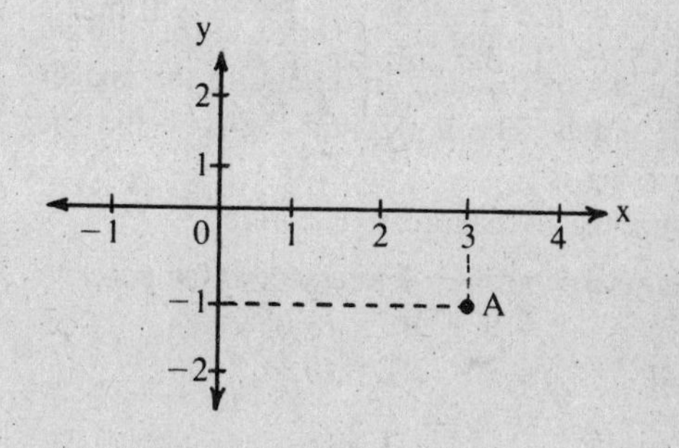

2. The diameter of a circle is a straight line passing through the center of the circle. The endpoints of the diameter are the same distance from the center.
   The center of the given circle is on the x-axis, at the origin. Point A is also on the x-axis, 4 units from the center. Point B must be on the x-axis, 4 units from the center.
   The coordinates of B are $(4,0)$.

   Answer: **(A)** $(4,0)$

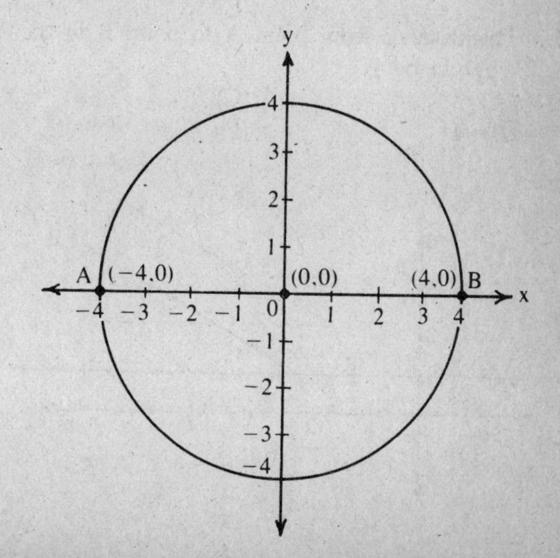

3. Substitute 1 for x in the equation:

$$y = 3 \cdot 1 - 4$$
$$= 3 - 4$$
$$= -1$$

Answer: **(A)** -1

4. The coordinates of the point of intersection must satisfy both equations. Choice B has x-coordinate = 7 and y-coordinate = 4.

Answer: **(B)** (7,4)

5. Point A has coordinates (2,1) and point B has coordinates (6,4). Using the distance formula,

$$d = \sqrt{(6-2)^2 + (4-1)^2}$$
$$= \sqrt{(4)^2 + (3)^2}$$
$$= \sqrt{16 + 9}$$
$$= \sqrt{25}$$
$$= 5$$

An alternate solution is to consider the right triangle formed by AB and the lines of the graph paper, with the right angle vertex at (6,1). Find the lengths of the legs by counting the spaces on the graph. The horizontal leg is 4 and the vertical leg is 3.

Using the Pythagorean Theorem,

$$(AB)^2 = 3^2 + 4^2$$
$$= 9 + 16$$
$$= 25$$
$$AB = \sqrt{25} = 5$$

Answer: **(C)** 5

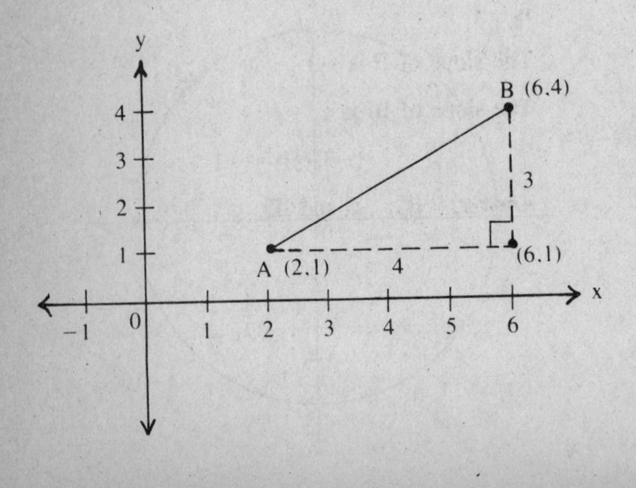

6. The line passes through the points (−2,0) and (0,2). Choice C is the only equation which is satisfied by both ordered pairs.

(−2,0): $y = x + 2$
$0 = -2 + 2$

(0,2): $y = x + 2$
$2 = 0 + 2$

Answer: **(C)** $y = x + 2$

7. Any point on the y-axis has its x-coordinate equal to 0. Substituting 0 for x in the equation,

$$x + 2y = 6$$
$$0 + 2y = 6$$
$$2y = 6$$
$$y = 3$$

Answer: **(A)** (0,3)

8. Graph both lines on the same set of axes. To find points on the graph of each line, choose any value for x and find the corresponding y.

For $y = 2x$:
if $x = 0$, $y = 2 \cdot 0 = 0$ (0,0)
if $x = 3$, $y = 2 \cdot 3 = 6$ (3,6)
if $x = -1$, $y = 2(-1) = -2$ (−1,−2)

For $y = x + 1$:
if $x = -1$, $y = -1 + 1 = 0$ (−1,0)
if $x = 2$, $y = 2 + 1 = 3$ (2,3)
if $x = 0$, $y = 0 + 1 = 1$ (0,1)

The point of intersection of the two lines is (1,2).

An alternate solution is to determine which of the given choices satisfies both equations.

Answer: **(A)** (1,2)

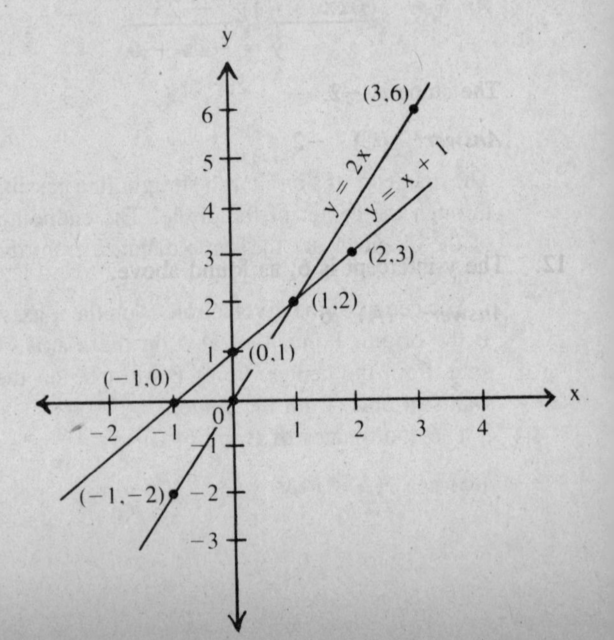

9. Using the distance formula,

$$d = \sqrt{(x_1 - x_2)^2 + (y_1 - y_2)^2}$$
$$d = \sqrt{[(-1) - 5]^2 + [0 - (-2)]^2}$$
$$= \sqrt{(-6)^2 + (+2)^2}$$
$$= \sqrt{36 + 4}$$
$$= \sqrt{40}$$
$$= \sqrt{4}\sqrt{10}$$
$$= 2\sqrt{10}$$

Answer: **(D)** $2\sqrt{10}$

10. Base AB = 10
Height = 3
Area = $\frac{1}{2}$(base)(height)
$= \frac{1}{2} \cdot 10 \cdot 3$
$= 15$

Answer: **(A)** 15

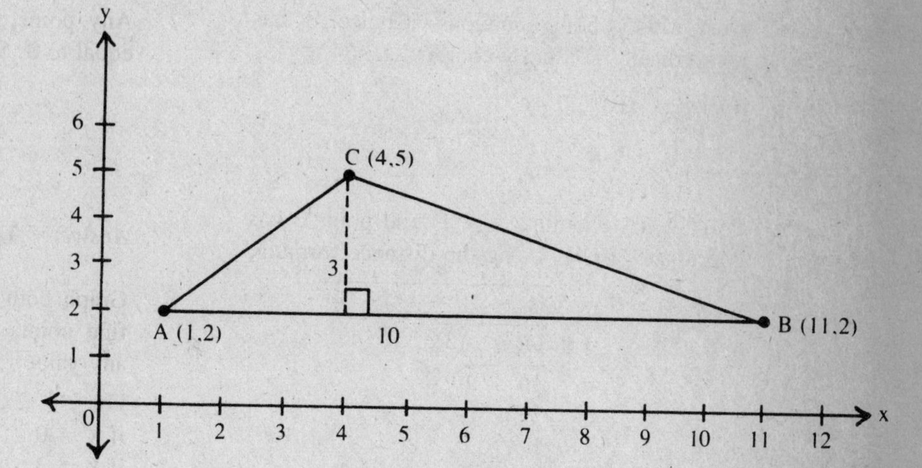

11. When a linear equation is written in the form $y = mx + b$, m is the slope and b is the y-intercept.

$$\begin{array}{r} 2x + y = 6 \\ -2x \qquad - 2x \\ \hline y = -2x + 6 \end{array}$$

The slope is -2.

Answer: **(C)** -2

12. The y-intercept is 6, as found above.

Answer: **(A)** 6

13.
$$m = \frac{y_1 - y_2}{x_1 - x_2}$$
$$= \frac{0 - 5}{-3 - 3} = \frac{-5}{-6} = \frac{5}{6}$$

Answer: **(C)** $\frac{5}{6}$

14. Parallel lines have equal slopes. The slopes of I and IV are both 3.

Answer: **(C)** I and IV only

15. The product of the slopes of perpendicular lines is -1.

The slope of II is -3

The slope of III is $\frac{1}{3}$

$$(-3)(\tfrac{1}{3}) = -1$$

Answer: **(C)** II and III

Part Two

THREE EXAMINATIONS FOR REVIEW

Answer Sheet for Examination I
Civil Service Arithmetic

1 Ⓐ Ⓑ Ⓒ Ⓓ	11 Ⓐ Ⓑ Ⓒ Ⓓ	21 Ⓐ Ⓑ Ⓒ Ⓓ	31 Ⓐ Ⓑ Ⓒ Ⓓ	41 Ⓐ Ⓑ Ⓒ Ⓓ
2 Ⓐ Ⓑ Ⓒ Ⓓ	12 Ⓐ Ⓑ Ⓒ Ⓓ	22 Ⓐ Ⓑ Ⓒ Ⓓ	32 Ⓐ Ⓑ Ⓒ Ⓓ	42 Ⓐ Ⓑ Ⓒ Ⓓ
3 Ⓐ Ⓑ Ⓒ Ⓓ	13 Ⓐ Ⓑ Ⓒ Ⓓ	23 Ⓐ Ⓑ Ⓒ Ⓓ	33 Ⓐ Ⓑ Ⓒ Ⓓ	43 Ⓐ Ⓑ Ⓒ Ⓓ
4 Ⓐ Ⓑ Ⓒ Ⓓ	14 Ⓐ Ⓑ Ⓒ Ⓓ	24 Ⓐ Ⓑ Ⓒ Ⓓ	34 Ⓐ Ⓑ Ⓒ Ⓓ	44 Ⓐ Ⓑ Ⓒ Ⓓ
5 Ⓐ Ⓑ Ⓒ Ⓓ	15 Ⓐ Ⓑ Ⓒ Ⓓ	25 Ⓐ Ⓑ Ⓒ Ⓓ	35 Ⓐ Ⓑ Ⓒ Ⓓ	45 Ⓐ Ⓑ Ⓒ Ⓓ
6 Ⓐ Ⓑ Ⓒ Ⓓ	16 Ⓐ Ⓑ Ⓒ Ⓓ	26 Ⓐ Ⓑ Ⓒ Ⓓ	36 Ⓐ Ⓑ Ⓒ Ⓓ	46 Ⓐ Ⓑ Ⓒ Ⓓ
7 Ⓐ Ⓑ Ⓒ Ⓓ	17 Ⓐ Ⓑ Ⓒ Ⓓ	27 Ⓐ Ⓑ Ⓒ Ⓓ	37 Ⓐ Ⓑ Ⓒ Ⓓ	47 Ⓐ Ⓑ Ⓒ Ⓓ
8 Ⓐ Ⓑ Ⓒ Ⓓ	18 Ⓐ Ⓑ Ⓒ Ⓓ	28 Ⓐ Ⓑ Ⓒ Ⓓ	38 Ⓐ Ⓑ Ⓒ Ⓓ	48 Ⓐ Ⓑ Ⓒ Ⓓ
9 Ⓐ Ⓑ Ⓒ Ⓓ	19 Ⓐ Ⓑ Ⓒ Ⓓ	29 Ⓐ Ⓑ Ⓒ Ⓓ	39 Ⓐ Ⓑ Ⓒ Ⓓ	49 Ⓐ Ⓑ Ⓒ Ⓓ
10 Ⓐ Ⓑ Ⓒ Ⓓ	20 Ⓐ Ⓑ Ⓒ Ⓓ	30 Ⓐ Ⓑ Ⓒ Ⓓ	40 Ⓐ Ⓑ Ⓒ Ⓓ	50 Ⓐ Ⓑ Ⓒ Ⓓ

EXAMINATION I
CIVIL SERVICE ARITHMETIC

50 questions — 1 hour

This examination presents the kinds of problems likely to be asked on the arithmetic section of a variety of civil service examinations. It provides an excellent review of the basic mathematical operations which are necessary to succeed in many occupations and trades.

DIRECTIONS: Solve each problem using available space on the page or a piece of scratch paper for your calculations. Then, from the four choices offered, select the one which you have figured out to be correct. Mark the letter of your choice on the answer sheet provided. When you have finished the exam, check your work against the solutions that appear at the end of the exam.

1. A bag of nickels and dimes contains $11.50. If there are 73 dimes, how many nickels are there?
(A) 78
(B) 80
(C) 82
(D) 84

2. A shipment consists of 340 ten-foot pieces of conduit with a coupling on each piece. If the conduit weighs 0.85 lb per foot and each coupling weighs 0.15 lb, the total weight of the shipment is
(A) 340 lb
(B) 628 lb
(C) 2941 lb
(D) 3400 lb

3. A carton contains 9 dozen file folders. If a clerk removes 53 folders, how many are left in the carton?
(A) 37
(B) 44
(C) 55
(D) 62

4. What tax rate on a base of $4782 would yield $286.92?
(A) 6%
(B) $8\frac{1}{4}$%
(C) 12%
(D) $16\frac{2}{3}$%

5. A can type 500 form letters in five hours. B can type 400 of these form letters in five hours. If A and B are to work together, the number of hours it will take them to type 540 form letters is most nearly
(A) 2
(B) 3
(C) 4
(D) 5

6. The difference between one tenth of 2000 and one-tenth percent of 2000 is
(A) 0
(B) 18
(C) 180
(D) 198

7. If the fractions $\frac{5}{7}$, $\frac{1}{2}$, $\frac{3}{5}$, and $\frac{2}{3}$ are arranged in ascending order of size, the result is
(A) $\frac{1}{2}, \frac{3}{5}, \frac{2}{3}, \frac{5}{7}$
(B) $\frac{3}{5}, \frac{5}{7}, \frac{2}{3}, \frac{1}{2}$
(C) $\frac{1}{2}, \frac{2}{3}, \frac{3}{5}, \frac{5}{7}$
(D) $\frac{5}{7}, \frac{2}{3}, \frac{3}{5}, \frac{1}{2}$

8. An employee has $\frac{2}{9}$ of his salary withheld for income tax. The percent of his salary that is withheld is most nearly
(A) 16%
(B) 18%
(C) 20%
(D) 22%

9. Frank and John repaired an old auto and sold it for $900. Frank worked on it 10 days and John worked 8 days. They divided the money in the ratio of the time spent on the work. Frank received
(A) $400
(B) $450
(C) $500
(D) $720

10. A driver traveled 100 miles at the rate of 40 mph, then traveled 80 miles at 60 mph. The total number of hours for the entire trip was
(A) $1\frac{3}{20}$
(B) $1\frac{3}{4}$
(C) $2\frac{1}{4}$
(D) $3\frac{5}{6}$

11. On a house plan on which 2 inches represents 5 feet, the length of a room measures $7\frac{1}{2}$ inches. The actual length of the room is
(A) $12\frac{1}{2}$ ft
(B) $15\frac{3}{4}$ ft
(C) $17\frac{1}{2}$ ft
(D) $18\frac{3}{4}$ ft

12. The ratio between .01% and .1 is
(A) 1 to 10
(B) 1 to 100
(C) 1 to 1000
(D) 1 to 10,000

13. After an article is discounted at 25%, it sells for $375. The original price of the article was
(A) $93.75
(B) $350
(C) $375
(D) $500

14. If Mr. Mitchell has $627.04 in his checking account and then writes three checks for $241.75, $13.24, and $102.97, what will be his new balance?
(A) $257.88 (C) $357.96
(B) $269.08 (D) $369.96

15. If erasers cost 8¢ each for the first 250, 7¢ each for the next 250, and 5¢ for every eraser thereafter, how many erasers may be purchased for $50?
(A) 600 (C) 850
(B) 750 (D) 1000

16. Assume that it is necessary to partition a room measuring 40 feet by 20 feet into eight smaller rooms of equal size. Allowing no room for aisles, the *minimum* amount of partitioning that would be needed is
(A) 90 ft (C) 110 ft
(B) 100 ft (D) 140 ft

17. As a result of reports received by the Housing Authority concerning the reputed ineligibility of 756 tenants because of above-standard incomes, an intensive check of their employers has been ordered. Four housing assistants have been assigned to this task. At the end of 6 days at 7 hours each, they have checked on 336 tenants. In order to speed up the investigation, two more housing assistants are assigned at this point. If they worked at the same rate, the number of additional 7-hour days it would take to complete the job is, most nearly,
(A) 1 (C) 5
(B) 3 (D) 7

18. A bird flying 400 miles covers the first 100 at the rate of 100 miles an hour, the second 100 at the rate of 200 miles an hour, the third 100 at the rate of 300 miles an hour, and the last 100 at the rate of 400 miles an hour. The average speed was
(A) 192 mph (C) 250 mph
(B) 212 mph (D) 150 mph

19. At 5 o'clock the smaller angle between the hands of the clock is
(A) 5° (C) 120°
(B) 75° (D) 150°

20. $7\frac{2}{3}\%$ of $1200 is
(A) $87 (C) $112
(B) $92 (D) $920

21. A certain family spends 30% of its income for food, 8% for clothing, 25% for shelter, 4% for recreation, 13% for education, and 5% for miscellaneous items. The weekly earnings are $500. What is the number of weeks it would take this family to save $15,000?
(A) 100 (C) 175
(B) 150 (D) 200

22. A 12-gallon mixture of antifreeze and water is 25% antifreeze. If 3 gallons of water are added to it, the strength of the mixture is now
(A) 12% (C) 20%
(B) $16\frac{2}{3}\%$ (D) 35%

23. A cab driver works on a commission basis, receiving $42\frac{1}{2}\%$ of the fares. In addition, his earnings from tips are valued at 29% of the commissions. If his average weekly fares equal $520, then his monthly earnings are
(A) between $900 and $1000
(B) between $1000 and $1100
(C) between $1100 and $1200
(D) over $1200

Questions 24 and 25 refer to the following graph:

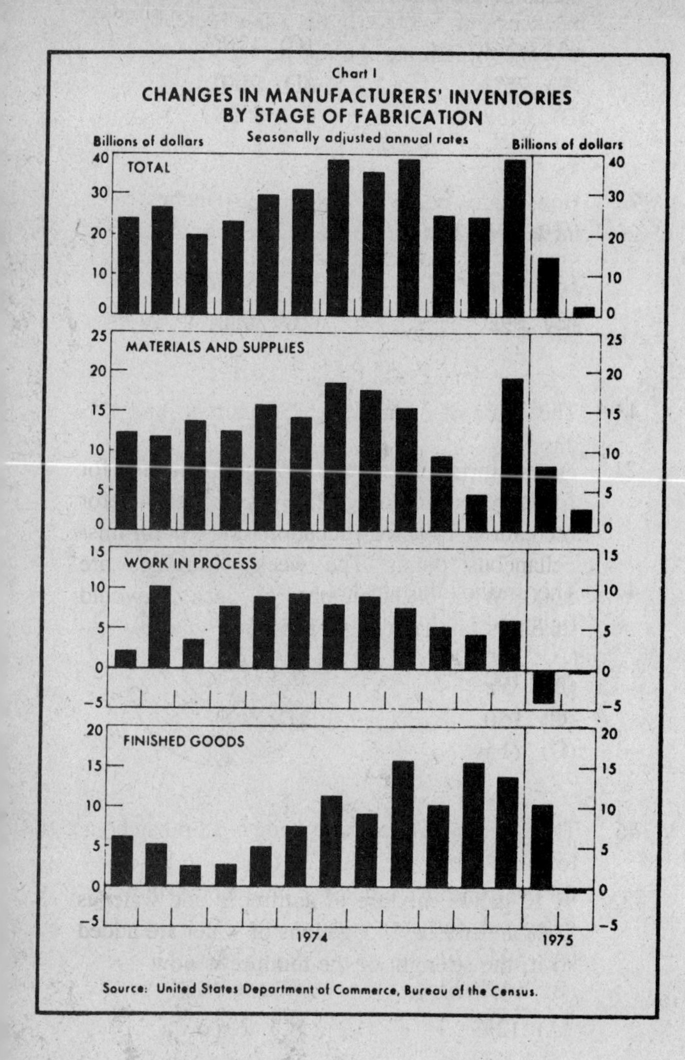

24. For how many months are materials and supplies inventories over $15 billion at the same time that finished goods inventories are over $10 billion?
    (A) one (C) three
    (B) two (D) four

25. In June 1974 the ratio of finished goods to work in process was approximately
    (A) 7:16 (C) 9:7
    (B) 9:16 (D) 7:9

26. An employee's net pay is equal to his total earnings less all deductions. If an employee's total earnings in a pay period are $497.05, what is his net pay if he has the following deductions: federal income tax, $90.32; FICA, $28.74; state tax, $18.79; city tax, $7.25; pension, $1.88?
    (A) $351.17 (C) $350.17
    (B) $351.07 (D) $350.07

27. Assume that two types of files have been ordered: 200 of type A and 100 of type B. When the files are delivered, the buyer discovers that 25% of each type is damaged. Of the remaining files, 20% of type A and 40% of type B are the wrong color. The total number of files that are the wrong color is
    (A) 30 (C) 50
    (B) 40 (D) 60

28. A parade is marching up an avenue for 60 city blocks. A sample count of the number of people watching the parade is taken, first in a block near the end of the parade, and then in a block at the middle. The former count is 4000, the latter is 6000. If the average for the entire parade is assumed to be the average of the two samples, then the estimated number of persons watching the entire parade is most nearly
    (A) 240,000 (C) 480,000
    (B) 300,000 (D) 600,000

29. If A takes 6 days to do a task and B takes 3 days to do the same task, working together they should do the same task in
    (A) $2\frac{2}{3}$ days (C) $2\frac{1}{3}$ days
    (B) 2 days (D) $2\frac{1}{2}$ days

30. The total length of fencing needed to enclose a rectangular area 46 feet by 34 feet is
    (A) 26 yd 1 ft (C) 52 yd 2 ft
    (B) $26\frac{2}{3}$ yd (D) $53\frac{1}{3}$ yd

31. Find the length of time it would take $432 to yield $74.52 in interest at $5\frac{3}{4}$% per annum.
    (A) 2 yr 10 mo (C) 3 yr 10 mo
    (B) 3 yr (D) 4 yr

32. The price of a radio is $31.29, which includes a 5% sales tax. What was the price of the radio before the tax was added?
    (A) $29.80 (C) $29.90
    (B) $29.85 (D) $29.95

33. If a person in the 19% income tax bracket pays $3515 in income taxes, his taxable income was
(A) $18,500
(B) $32,763
(C) $53,800
(D) $67,785

34. Of the numbers 6, 5, 3, 3, 6, 3, 4, 3, 4, 3, the mode is
(A) 3
(B) 4
(C) 5
(D) 6

35. In a circle graph a sector of 108 degrees is shaded to indicate the overhead in doing $150,000 gross business. The overhead amounts to
(A) $1200
(B) $4500
(C) $12,000
(D) $45,000

36. Two people start at the same point and walk in opposite directions. If one walks at the rate of 2 miles per hour and the other walks at the rate of 3 miles per hour, in how many hours will they be 20 miles apart?
(A) 2
(B) 3
(C) 4
(D) 5

37. In a group of 100 people, 37 wear glasses. What is the probability that a person chosen at random from this group does *not* wear glasses?
(A) .37
(B) .50
(C) .63
(D) 1.00

38. The interest on $148.00 at 6% for 60 days is
(A) $8.88
(B) $2.96
(C) $14.80
(D) $ 1.48

39. A man bought a camera that was listed at $160. He was given successive discounts of 20% and 10%. The price he paid was
(A) $112.00
(B) $115.20
(C) $119.60
(D) $129.60

40. The water level of a swimming pool measuring 75 feet by 42 feet is to be raised four inches. If there are 7.48 gallons in a cubic foot, the number of gallons of water that will be needed is
(A) 140
(B) 31,500
(C) 7854
(D) 94,500

41. A salesman is paid $4\frac{1}{2}$% commission on his first $7000 of sales and $5\frac{1}{2}$% commission on all sales in excess of $7000. If his sales were $9600, how much commission did he earn?
(A) $432
(B) $458
(C) $480
(D) $528

42. How many boxes 3 inches by 4 inches by 5 inches can fit into a carton 3 feet by 4 feet by 5 feet?
(A) 60
(B) 144
(C) 1728
(D) 8640

43. The value of 32 nickels, 73 quarters, and 156 dimes is
(A) $26.10
(B) $31.75
(C) $35.45
(D) $49.85

44. The area of the shaded figure is
(A) 4π
(B) 5π
(C) 16π
(D) 21π

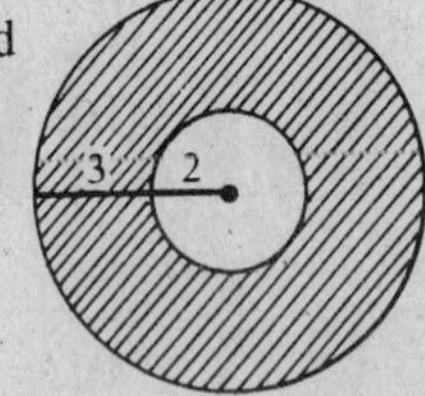

45. The wage rate in a certain trade is $8.60 an hour for a 40-hour week and $1\frac{1}{2}$ times the base pay for overtime. An employee who works 48 hours in a week earns
(A) $447.20
(B) $498.20
(C) $582.20
(D) $619.20

46. Jane Michaels borrowed $200 on March 31 at the simple interest rate of 8% per year. If she wishes to repay the loan and the interest on May 15, what is the total amount she must pay?
(A) $201
(B) $202
(C) $203
(D) $204

47. How many decigrams are in .57 kilograms?
(A) 57
(B) 570
(C) 5700
(D) 57,000

48. If candies are bought at $1.10 per dozen and sold at 3 for 55 cents, the total profit on $5\frac{1}{2}$ dozen is
(A) $5.55
(B) $6.05
(C) $6.55
(D) $7.05

49. The number missing in the sequence 2, 5, 10, 17, ——, 37, 50, 65 is
(A) 22 (C) 26
(B) 24 (D) 27

50. A cylindrical container has a diameter of 14 inches and a height of 6 inches. If one gallon equals 231 cubic inches, the capacity of the tank is approximately
(A) $2\frac{2}{7}$ gal (C) $1\frac{1}{7}$ gal
(B) 4 gal (D) 3 gal

Answer Key for Examination I Civil Service Arithmetic

1. **(D)**	11. **(D)**	21. **(D)**	31. **(B)**	41. **(B)**
2. **(C)**	12. **(C)**	22. **(C)**	32. **(A)**	42. **(C)**
3. **(C)**	13. **(D)**	23. **(C)**	33. **(A)**	43. **(C)**
4. **(A)**	14. **(B)**	24. **(C)**	34. **(A)**	44. **(D)**
5. **(B)**	15. **(B)**	25. **(D)**	35. **(D)**	45. **(A)**
6. **(D)**	16. **(B)**	26. **(D)**	36. **(C)**	46. **(B)**
7. **(A)**	17. **(C)**	27. **(D)**	37. **(C)**	47. **(C)**
8. **(D)**	18. **(A)**	28. **(B)**	38. **(D)**	48. **(B)**
9. **(C)**	19. **(D)**	29. **(B)**	39. **(B)**	49. **(C)**
10. **(D)**	20. **(B)**	30. **(D)**	40. **(C)**	50. **(B)**

Solutions for Examination I

1. 73 dimes = 73 × \$.10
= \$7.30
\$11.50 − \$7.30 = \$4.20

There is \$4.20 worth of nickels in the bag.

\$4.20 ÷ \$.05 = 84 nickels

Answer: **(D)** 84

2. Each 10-foot piece weighs

10 × .85 lb = 8.5 lb
+ .15 lb
8.65 lb

The entire shipment weighs

340 × 8.65 lb = 2941 lb

Answer: **(C)** 2941 lb

3. The carton contains 9 × 12 = 108 folders.

108 − 53 = 55 remain in carton

Answer: **(C)** 55

4. Rate = tax ÷ base

286.92 ÷ 4782 = 4782) 286.92 = .06
286 92

.06 = 6%

Answer: **(A)** 6%

5. A can type $500 \div 5 = 100$ letters per hour.
B can type $400 \div 5 = 80$ letters per hour.

Together they type 180 letters per hour.

$$540 \div 180 = 3$$

It will take 3 hours to type 540 letters.

Answer: **(B)** 3

6. $$\tfrac{1}{10} \text{ of } 2000 = \tfrac{1}{10} \times 2000 = 200$$
$$\tfrac{1}{10}\% \text{ of } 2000 = .001 \times 2000 = 2$$

The difference is $200 - 2 = 198$.

Answer: **(D)** 198

7. To compare the fractions, change them to fractions having the same denominator.

$$\text{L.C.D.} = 7 \times 2 \times 5 \times 3 = 210$$
$$\tfrac{5}{7} = \tfrac{150}{210}$$
$$\tfrac{1}{2} = \tfrac{105}{210}$$
$$\tfrac{3}{5} = \tfrac{126}{210}$$
$$\tfrac{2}{3} = \tfrac{140}{210}$$

The correct order is $\tfrac{105}{210}, \tfrac{126}{210}, \tfrac{140}{210}, \tfrac{150}{210}$.

Answer: **(A)** $\tfrac{1}{2}, \tfrac{3}{5}, \tfrac{2}{3}, \tfrac{5}{7}$

8. $$\tfrac{2}{9} = 9\overline{)2.00} = .22\tfrac{2}{9}$$

```
      .22 2/9
 9 ) 2.00
     1 8
     ---
       20
       18
       --
        2
```

$.22\tfrac{2}{9} = 22\%$ approximately

Answer: **(D)** 22%

9. Frank and John worked in the ratio 10:8.

$$10 + 8 = 18$$
$$\$900 \div 18 = \$50$$
$$\text{Frank's share} = 10 \times \$50 = \$500$$

Answer: **(C)** $500

10. The first part of the trip took
$$100 \text{ mi} \div 40 \text{ mph} = 2\tfrac{1}{2} \text{ hours}$$
The second part of the trip took
$$80 \text{ mi} \div 60 \text{ mph} = 1\tfrac{1}{3} \text{ hours}$$

$$\begin{array}{rcr} 2\tfrac{1}{2} & = & 2\tfrac{3}{6} \\ +\ 1\tfrac{1}{3} & = & +\ 1\tfrac{2}{6} \\ \hline & & 3\tfrac{5}{6} \end{array}$$

Answer: **(D)** $3\tfrac{5}{6}$

11. Let f represent the actual number of feet. The plan lengths and the actual lengths are in proportion. Therefore,

$$\frac{2}{7\frac{1}{2}} = \frac{5}{f}$$

and
$$f = \frac{5 \times 7\frac{1}{2}}{2}$$
$$= \frac{5 \times \frac{15}{2}}{2}$$
$$= \tfrac{75}{2} \div 2$$
$$= \tfrac{75}{2} \times \tfrac{1}{2}$$
$$= \tfrac{75}{4} = 18\tfrac{3}{4}$$

Answer: **(D)** $18\tfrac{3}{4}$ ft

12. $$\frac{.01\%}{.1} = \frac{.0001}{.1}$$
$$= .001$$
$$= \frac{1}{1000}$$

Answer: **(C)** 1 to 1000

13. $375 is 75% of the original price.

$$\text{The original price} = \$375 \div 75\%$$
$$= \$375 \div .75$$
$$= \$500$$

Answer: **(D)** $500

14. Total of checks:

$$\begin{array}{r} \$241.75 \\ 13.24 \\ +\ 102.97 \\ \hline \$357.96 \end{array}$$

$$\begin{array}{rl} \$627.04 & \text{old balance} \\ -\ 357.96 & \text{checks} \\ \hline \$269.08 & \text{new balance} \end{array}$$

Answer: **(B)** $269.08

15. First 250 erasers:
$$250 \times \$.08 = \$20.00$$
Next 250 erasers:
$$250 \times \$.07 = \$17.50$$
Total for 500 erasers:
$$\$20.00 + \$17.50 = \$37.50$$
$$\$50.00 - \$37.50 = \$12.50$$
$12.50 remains for 5¢ erasers:
$$\$12.50 \div \$.05 = 250 \text{ erasers}$$
$$500 + 250 = 750$$

Answer: **(B)** 750

16. The room may be partitioned as is shown below:

40
20
10 10 10
10 10 10 10
10 10 10

The total amount of partitioning is 100 feet.

Answer: **(B)** 100 ft

17. Four assistants completed 336 cases in 42 hours (6 days at 7 hours per day). Therefore, each assistant completed 336 ÷ 4, or 84 cases in 42 hours, for a rate of 2 cases per hour per assistant.

After the first 6 days, the number of cases remaining is

$$756 - 336 = 420$$

It will take 6 assistants, working at the rate of 2 cases per hour per assistant 420 ÷ 12 or 35 hours to complete the work. If each workday has 7 hours, then 35 ÷ 7 or 5 days are needed.

Answer: **(C)** 5

18. At 100 mph, 100 miles will take 1 hour. At 200 mph, 100 miles will take $\frac{1}{2}$ hour. At 300 mph, 100 miles will take $\frac{1}{3}$ hour. At 400 mph, 100 miles will take $\frac{1}{4}$ hour.

Total time:

$$\begin{aligned} 1 &= \tfrac{12}{12} \\ \tfrac{1}{2} &= \tfrac{6}{12} \\ \tfrac{1}{3} &= \tfrac{4}{12} \\ +\tfrac{1}{4} &= +\tfrac{3}{12} \\ \hline & \tfrac{25}{12} = 2\tfrac{1}{12} \text{ hours} \end{aligned}$$

$$400 \text{ miles} \div 2\tfrac{1}{12} \text{ hours} = 400 \div \tfrac{25}{12} \text{ mph}$$

$$= \overset{16}{\cancel{400}} \times \frac{12}{\underset{1}{\cancel{25}}} \text{ mph}$$

$$= 192 \text{ mph}$$

Answer: **(A)** 192 mph

19. Each hour is represented by

$$360° \div 12 = 30°$$

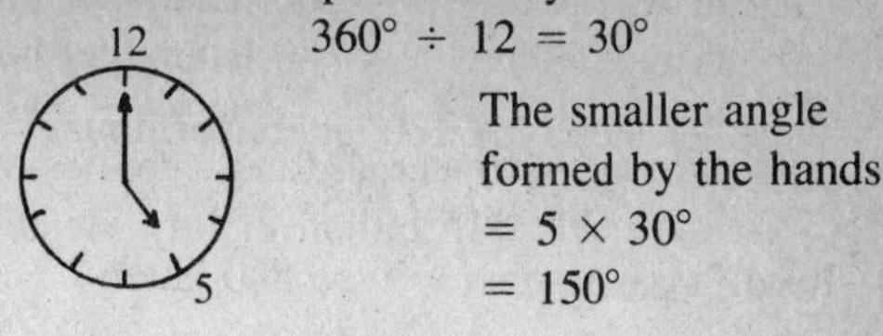

The smaller angle formed by the hands $= 5 \times 30°$ $= 150°$

Answer: **(D)** 150°

20. $7\frac{2}{3}\%$ of \$1200 $= \frac{23}{3}\%$ of \$1200

To change $\frac{23}{3}\%$ to a fraction, divide by 100:

$$\frac{23}{3} \div 100 = \frac{23}{3} \times \frac{1}{100} = \frac{23}{300}$$

$$\frac{23}{\underset{1}{\cancel{300}}} \times \overset{4}{\cancel{\$1200}} = \$92$$

Answer: **(B)** \$92

21. The family spends a total of 85% of its income. Therefore, 100% − 85%, or 15%, remains for savings.

$$\begin{aligned} 15\% \text{ of } \$500 &= .15 \times \$500 \\ &= \$75 \text{ per week} \\ \$15{,}000 \div \$75 &= 200 \text{ weeks} \end{aligned}$$

Answer: **(D)** 200

22. The mixture contains 25% of 12 gallons, or 3 gallons, of antifreeze. The remaining 9 gallons must be water.

The new mixture would contain 3 gallons of antifreeze and 9 + 3 = 12 gallons of water, for a total of 15 gallons. The strength would be

$$\frac{3 \text{ gal. antifreeze}}{15 \text{ gal. mixture}} = .20$$

$$= 20\%$$

Answer: **(C)** 20%

23. Commission = $42\frac{1}{2}\%$ of fares

$$\begin{aligned} 42\tfrac{1}{2}\% \text{ of } \$520 &= .425 \times \$520 \\ &= \$221 \text{ commission} \end{aligned}$$

Tips = 29% of commission

$$\begin{aligned} 29\% \text{ of } \$221 &= .29 \times \$221 \\ &= \$64.09 \text{ tips} \end{aligned}$$

Weekly earnings:

$$\begin{array}{r} \$221.00 \\ +\quad 64.09 \\ \hline \$285.09 \end{array}$$

Monthly earnings:

$$\begin{array}{r} \$285.09 \\ \times \quad 4 \\ \hline \$1140.36 \end{array}$$

Answer: **(C)** between \$1100 and \$1200

24. Finished goods inventories are over \$10 billion in July, September, October, November, and December of 1974 and January of 1975. Of those months, materials and supplies inventories are over \$15 billion in July, September, and December.

Answer: **(C)** three

25. In June 1974 finished goods inventories were approximately \$7 billion, and work in process inventories were approximately \$9 billion. The ratio is 7:9.

Answer: **(D)** 7:9

26.

	\$ 90.32
	28.74
	18.79
	7.25
	1.88
Total deductions	\$146.98

\$497.05	total earnings
− 146.98	deductions
\$350.07	net pay

Answer: **(D)** \$350.07

27. If 25% are damaged, then 75% are not damaged.

Type A: $75\% \text{ of } 200 = .75 \times 200 = 150$

20% of 150 are wrong color

$20\% \text{ of } 150 = .20 \times 150 = 30$

Type B: $75\% \text{ of } 100 = .75 \times 100 = 75$

40% of 75 are wrong color

$40\% \text{ of } 75 = .40 \times 75 = 30$

Total wrong color = 30 + 30 = 60

Answer: **(D)** 60

28. Average is $\frac{4000 + 6000}{2} = 5000$ per block.

If there are 60 blocks, there are

$60 \times 5000 = 300{,}000$ people

Answer: **(B)** 300,000

29. A can do $\frac{1}{6}$ of the task in 1 day, and B can do $\frac{1}{3}$ in 1 day.

Together, in 1 day they can do

$$\begin{array}{r} \frac{1}{6} = \frac{1}{6} \\ +\frac{1}{3} = +\frac{2}{6} \\ \hline \frac{3}{6} = \frac{1}{2} \text{ of the job} \end{array}$$

It will take 2 days to complete the job if they work together.

Answer: **(B)** 2 days

30.

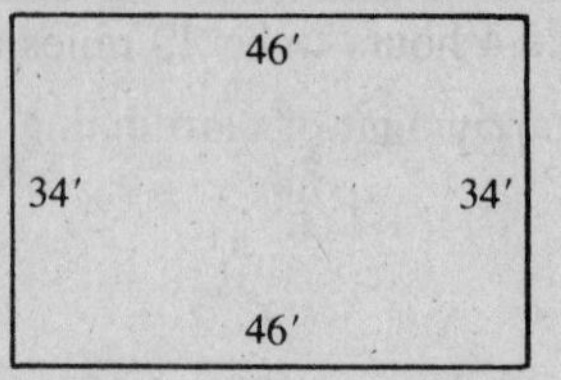

$$\text{Perimeter} = 46' + 34' + 46' + 34' = 160'$$

$$160 \text{ ft} \div 3 \text{ ft per yd} = \frac{160}{3} \text{ yd} = 53\frac{1}{3} \text{ yd}$$

Answer: **(D)** $53\frac{1}{3}$ yd

31. $$\$432 \times 5\frac{3}{4}\% = \overset{108}{\cancel{432}} \times \frac{23}{\underset{100}{\cancel{400}}} = \$\frac{2484}{100} = \$24.84$$

$$\$74.52 \div \$24.84 = 3$$

Answer: **(B)** 3 yr

32. \$31.29 = 105% of price before tax

Price before tax = \$31.29 ÷ 105%
= \$31.29 ÷ 1.05
= \$29.80

Answer: **(A)** \$29.80

33. \$3515 = 19% of taxable income

Taxable income = \$3515 ÷ 19%
= \$3515 ÷ .19
= \$18,500

Answer: **(A)** \$18,500

34. The mode is the value appearing most frequently. For the list given, the mode is 3.

Answer: **(A)** 3

35. A sector of 108° is

$$\frac{108°}{360°} = \frac{3}{10} \text{ of the circle}$$

$$\frac{3}{10} \times \$150{,}000 = \$45{,}000$$

Answer: **(D)** $45,000

36. In 1 hour they are 5 miles apart.

$$20 \text{ mi} \div 5 \text{ mi} = 4 \text{ hr}$$

It will take 4 hours to be 20 miles apart.

Answer: **(C)** 4

37. If 37 wear glasses, 100 − 37, or 63 do not wear glasses.

The probability is $\frac{63}{100} = .63$

Answer: **(C)** .63

38.

$$60 \text{ days} = \frac{60}{360} \text{ year} = \frac{1}{6} \text{ year}$$

$$\text{Interest} = \$148 \times .06 \times \frac{1}{6} = \$1.48$$

Answer: **(D)** $1.48

39. First discount:

20% of $160 = .20 × $160 = $32

$160 − $32 = $128

Second discount:

10% of $128 = .10 × $128 = $12.80

$128.00 − $12.80 = $115.20

Answer: **(B)** $115.20

40.

$$4 \text{ in} = \frac{1}{3} \text{ ft}$$

$$\text{Volume to be added} = \overset{25}{\cancel{75}} \times 42 \times \frac{1}{\cancel{3}_{1}}$$

$$= 1050 \text{ cu ft}$$

$$= 1050 \times 7.48 \text{ gal}$$

$$= 7854 \text{ gal}$$

Answer: **(C)** 7854

41. Commission on first $7000:

$4\frac{1}{2}\%$ of $7000 = .045 × $7000 = $315

Commission on remainder:

$9600 − $7000 = $2600

$5\frac{1}{2}\%$ of $2600 = .055 × $2600 = $143

Total commission = $315 + $143 = $458

Answer: **(B)** $458

42. Volume of the carton = 3 ft × 4 ft × 5 ft

= 36 in × 48 in × 60 in

= 103,680 cu in

Volume of each box = 3 in × 4 in × 5 in

= 60 cu in

103,680 ÷ 60 = 1728

Answer: **(C)** 1728

43.

32 nickels	=	32 × $.05 =	$ 1.60
73 quarters	=	73 × $.25 =	18.25
156 dimes	=	156 × $.10 =	15.60
		Total =	$35.45

Answer: **(C)** $35.45

44. The area of the shaded figure equals the area of the larger circle minus the area of the smaller circle.

Area of larger circle = $5^2\pi = 25\pi$

− Area of smaller circle = $2^2\pi = 4\pi$

Area of shaded figure = 21π

Answer: **(D)** 21π

45.

48 − 40 = 8 hours overtime

Salary for 8 hours overtime:

$$1\tfrac{1}{2} \times \$8.60 \times 8 = \frac{3}{\cancel{2}_{1}} \times \$8.60 \times \overset{4}{\cancel{8}}$$

$$= \$103.20$$

Salary for 40 hours regular time:

$8.60 × 40 = $344.00

Total salary = $344.00 + $103.20 = $447.20

Answer: **(A)** $447.20

46. From March 31 to May 15 is 45 days, which is $\frac{45}{360}$ of a year.

$$\text{Interest} = \$200 \times .08 \times \frac{\overset{1}{\cancel{45}}}{\underset{8}{\cancel{360}}}$$
$$= \$\tfrac{16}{8}$$
$$= \$2$$

She must pay $200 + $2 = $202.

Answer: **(B)** $202

47. $$.57 \text{ kilograms} = .57 \times 1000 \text{ grams}$$
$$= 570 \text{ grams}$$
$$= 570 \div .10 \text{ decigrams}$$
$$= 5700 \text{ decigrams}$$

Answer: **(C)** 5700

48. The cost of $5\frac{1}{2}$ dozen is
$$5\tfrac{1}{2} \times \$1.10 = 5.5 \times \$1.10$$
$$= \$6.05$$

The candies sell at 3 for $.55. A dozen sell for 4 × $.55, or $2.20. The selling price of $5\frac{1}{2}$ dozen is
$$5\tfrac{1}{2} \times \$2.20 = 5.5 \times \$2.20$$
$$= \$12.10$$
$$\text{Profit} = \$12.10 - \$6.05$$
$$= \$6.05$$

Answer: **(B)** $6.05

49. Find the differences between terms:

2 ‿ 5 ‿ 10 ‿ 17 —— 37 ‿ 50 ‿ 65
 3   5   7   13   15

The difference between 17 and the missing term must be 9. The missing term is
$$17 + 9 = 26$$

Answer: **(C)** 26

50. The volume of a cylinder = πr^2h. If the diameter is 14, the radius is 7. Using $\pi = \frac{22}{7}$,
$$\text{Volume} = \frac{22}{\underset{1}{\cancel{7}}} \times \overset{7}{\cancel{49}} \times 6$$
$$= 924 \text{ cubic inches}$$
$$924 \div 231 = 4 \text{ gallons}$$

Answer: **(B)** 4 gal

Answer Sheet for Examination II
Mathematics for GED Candidates

1 ① ② ③ ④ ⑤	11 ① ② ③ ④ ⑤	21 ① ② ③ ④ ⑤	31 ① ② ③ ④ ⑤	41 ① ② ③ ④ ⑤
2 ① ② ③ ④ ⑤	12 ① ② ③ ④ ⑤	22 ① ② ③ ④ ⑤	32 ① ② ③ ④ ⑤	42 ① ② ③ ④ ⑤
3 ① ② ③ ④ ⑤	13 ① ② ③ ④ ⑤	23 ① ② ③ ④ ⑤	33 ① ② ③ ④ ⑤	43 ① ② ③ ④ ⑤
4 ① ② ③ ④ ⑤	14 ① ② ③ ④ ⑤	24 ① ② ③ ④ ⑤	34 ① ② ③ ④ ⑤	44 ① ② ③ ④ ⑤
5 ① ② ③ ④ ⑤	15 ① ② ③ ④ ⑤	25 ① ② ③ ④ ⑤	35 ① ② ③ ④ ⑤	45 ① ② ③ ④ ⑤
6 ① ② ③ ④ ⑤	16 ① ② ③ ④ ⑤	26 ① ② ③ ④ ⑤	36 ① ② ③ ④ ⑤	46 ① ② ③ ④ ⑤
7 ① ② ③ ④ ⑤	17 ① ② ③ ④ ⑤	27 ① ② ③ ④ ⑤	37 ① ② ③ ④ ⑤	47 ① ② ③ ④ ⑤
8 ① ② ③ ④ ⑤	18 ① ② ③ ④ ⑤	28 ① ② ③ ④ ⑤	38 ① ② ③ ④ ⑤	48 ① ② ③ ④ ⑤
9 ① ② ③ ④ ⑤	19 ① ② ③ ④ ⑤	29 ① ② ③ ④ ⑤	39 ① ② ③ ④ ⑤	49 ① ② ③ ④ ⑤
10 ① ② ③ ④ ⑤	20 ① ② ③ ④ ⑤	30 ① ② ③ ④ ⑤	40 ① ② ③ ④ ⑤	50 ① ② ③ ④ ⑤

EXAMINATION II MATHEMATICS FOR GED CANDIDATES

50 questions — 90 minutes

This examination is patterned on the mathematics section of the High School Equivalency Diploma Test. The problems are very similar to the ones on the actual test. The time allowed and number of questions is just what GED candidates can expect.

DIRECTIONS: Study each of the following problems and work out your answers in the margins or on a piece of scratch paper. Below each problem you will find five suggested answers, numbered from 1 to 5. Select the answer you have figured out to be correct and mark its number on the answer sheet provided. Solutions for each problem appear at the end of the examination.

1. The difference between three hundred four thousand eight hundred two and two hundred twelve thousand eight hundred ten is

(1) 91,992
(2) 96,592
(3) 182,328
(4) 209,328
(5) 210,972

2. Joan earns $4.00 per hour. On a day that she works from 9:30 AM to 3:00 PM, how much will she earn?

(1) $14.00
(2) $18.00
(3) $22.00
(4) $26.00
(5) $30.00

3. The product of .010 and .001 is

(1) .01100
(2) .10100
(3) .00001
(4) .01000
(5) .01010

4. If $2x + y = 7$, what is the value of y when $x = 3$?

(1) 1
(2) 3
(3) 5
(4) 7
(5) 9

5. Paul received a bonus of $750, which was 5% of his annual salary. His annual salary was

(1) $37,500
(2) $25,000
(3) $22,500
(4) $15,000
(5) $7,500

6. The value of $(-6) + (-2)(-3)$ is

(1) -24
(2) -12
(3) 0
(4) 12
(5) 24

7. Round 825.6347 to the nearest hundredth.

(1) 800
(2) 825.63
(3) 825.64
(4) 825.635
(5) 825.645

8. The coordinates of point P on the graph are

y
P
1
0
1

(1) $(2,-3)$
(2) $(-3,2)$
(3) $(-2,3)$
(4) $(3,-2)$
(5) $(-2,-3)$

9. A boy buys oranges at 3 for 30¢ and sells them at 5 for 60¢. How many oranges must he sell in order to make a profit of 50¢?

(1) 12
(2) 25
(3) 50
(4) 75
(5) 100

10. The formula for the volume of a right circular cone is $V = \frac{1}{3}\pi r^2 h$, where r is the radius and h is the height. Find the approximate volume of a right circular cone which has radius 3 inches and height 14 inches (π is approximately $\frac{22}{7}$).

(1) 33 cubic inches
(2) 132 cubic inches
(3) 396 cubic inches
(4) 686 cubic inches
(5) 1188 cubic inches

11. Of the following, the number which is nearest in value to 5 is

(1) 4.985
(2) 5.005
(3) 5.01
(4) 5.1
(5) 5.105

12. If a rope four yards long is cut into three equal pieces, how long will each piece be?

(1) 4 feet
(2) $3\frac{1}{2}$ feet
(3) $3\frac{1}{3}$ feet
(4) 3 feet
(5) $2\frac{1}{4}$ feet

13. The number of square units in the area of triangle ABC is

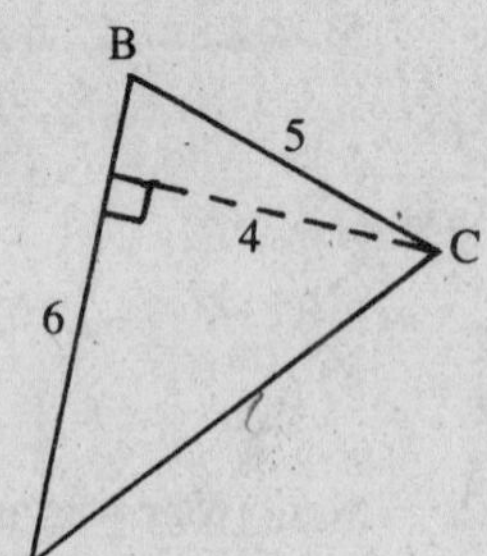

(1) 35
(2) 28
(3) 24
(4) 14
(5) 12

14. The cost of 30 sandwich rolls at $1.50 per dozen is

(1) $3.00
(2) $3.45
(3) $3.75
(4) $4.50
(5) $4.80

15. If $P = 2(a + b)$, find P when $a = 3$ and $b = 4$.

(1) 9
(2) 10
(3) 14
(4) 24
(5) 28

16. Two angles of a triangle measure 30° and 50°. The number of degrees in the third angle is

(1) 10
(2) 40
(3) 50
(4) 90
(5) 100

17. A map is drawn to the scale $1\frac{1}{2}$ inches = 50 miles. What is the actual distance between two towns which are $4\frac{1}{2}$ inches apart on the map?

(1) 45 miles
(2) 90 miles
(3) 120 miles
(4) 150 miles
(5) 300 miles

18. The numbers in the sequence 1, 4, 9, 16, 25, . . . follow a particular pattern. If the pattern is continued, what number should appear after 25?

(1) 28
(2) 30
(3) 34
(4) 36
(5) 40

19. If shipping charges to a certain point are $1.24 for the first five ounces and 16 cents for each additional ounce, the weight of a package for which the charges are $3.32 is

(1) 13 ounces
(2) 15 ounces
(3) $1\frac{1}{8}$ pounds
(4) $1\frac{1}{4}$ pounds
(5) $1\frac{1}{2}$ pounds

20. If a recipe for a cake calls for $2\frac{1}{2}$ cups of flour, and Mary wishes to make three such cakes, the number of cups of flour she must use is

(1) 5
(2) $6\frac{1}{2}$
(3) $7\frac{1}{2}$
(4) 9
(5) $9\frac{1}{2}$

21. The equation of the line passing through the points (−2,2) and (3,−3) is

(1) $x + y = 5$
(2) $x - y = 5$
(3) $y - x = 5$
(4) $y = x$
(5) $y = -x$

22. What will it cost to carpet a room 12 feet wide and 15 feet long if carpeting costs $20.80 per square yard?

(1) $334.60
(2) $374.40
(3) $416.00
(4) $504.60
(5) $560.00

23. If a five pound mixture of nuts contains two pounds of cashews and the rest peanuts, what percent of the mixture is peanuts?

(1) 20
(2) 30
(3) 40
(4) 50
(5) 60

Questions 24–26 refer to the graph below:

RAINFALL IN DAMP CITY
January–July, 1980

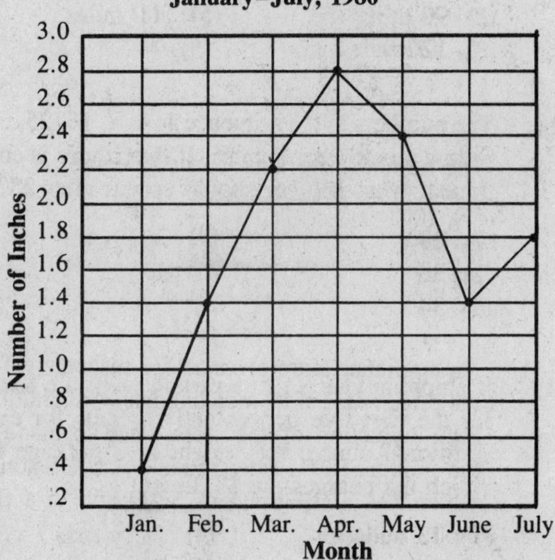

24. The total rainfall for the months January, February and March was, in inches,

(1) 2.2
(2) 3.4
(3) 4.0
(4) 4.8
(5) 7.6

25. The average monthly rainfall, in inches, for April, May and June was

(1) 2.0
(2) 2.1
(3) 2.2
(4) 2.3
(5) 2.4

26. Which statement about the information given in the graph is *false*?

(1) The rainfall in April was twice the rainfall in February.
(2) June had greater rainfall than February.
(3) The month with the least rainfall was January.
(4) March had .4 inches greater rainfall than July.
(5) May had more rain than March.

27. How long will the shadow of a five foot tall person be at the same time that an eight foot high pole casts a shadow 24 feet long?

(1) 1 foot
(2) 8 feet
(3) 15 feet
(4) $32\frac{1}{2}$ feet
(5) 72 feet

28. George has a five dollar bill and a ten dollar bill. If he buys one item costing $7.32 and another item costing $1.68, how much money will he have left?

(1) $1.10
(2) $5.64
(3) $6.00
(4) $9.00
(5) $9.90

29. If one card is picked at random from a deck of cards, the probability that it is a club is

(1) 1
(2) $\frac{1}{52}$
(3) $\frac{1}{13}$
(4) $\frac{1}{10}$
(5) $\frac{1}{4}$

30. Jack can ride his bicycle 6 miles in 48 minutes. At the same rate, how long will it take him to ride 15 miles?

(1) 1 hour 20 minutes
(2) 2 hours
(3) 2 hours 12 minutes
(4) 3 hours
(5) 3 hours 12 minutes

31. If $2x - 7 = 3$, then $3x + 1 =$

(1) 4
(2) 5
(3) 7
(4) 12
(5) 16

32. Over a four year period, the sales of the Acme Company increased from $13,382,675 to $17,394,683. The average yearly increase was

(1) $4,012,008
(2) $3,146,014
(3) $2,869,054
(4) $1,060,252
(5) $1,003,002

33. The perimeter of figure ABCDE is

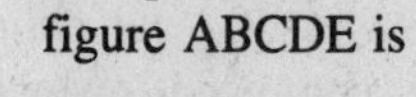

(1) 18
(2) 25
(3) 38
(4) 44
(5) 45

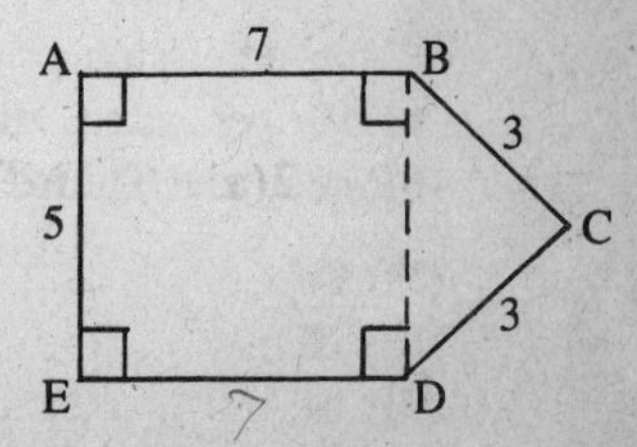

34. Of the following, the unit which would most likely be used to measure the distance from New York to Albany is the

(1) liter
(2) kilometer
(3) centigram
(4) millimeter
(5) degree Celsius

35. The simple interest on \$200 at 12% for 2 years is

(1) \$6
(2) \$12
(3) \$24
(4) \$48
(5) \$120

Questions 36–38 refer to the picture graph below. The picture graph represents how many men ( ), women ( ), boys (○), and girls (△) visited a museum one particular week. Each figure represents 100.

Mon.	*Tues.*	*Wed.*	*Thurs.*	*Fri.*
Man	Man	Man	Woman	Woman
Man	Woman	Woman	Woman	Woman
Woman	△	Woman	△	Man
○	△	○	△	Man
○	○	△	○	○
△		△	○	△
			○	

36. Over the five-day period, the ratio of men visitors to women visitors was

(1) 3:4
(2) 4:3
(3) 3:7
(4) 4:7
(5) 7:3

37. If the admission price was 50¢ per child and \$1.50 per adult, the combined revenue on Monday and Thursday was

(1) \$11.50
(2) \$260
(3) \$1150
(4) \$2600
(5) \$11,500

38. The total number of visitors to the museum during the week was

(1) 3000
(2) 2200
(3) 1400
(4) 300
(5) 30

39. A man spent exactly one dollar in the purchase of 3-cent stamps and 5-cent stamps. The number of 5-cent stamps which he could *not* have purchased under the circumstances is

(1) 5
(2) 8
(3) 9
(4) 11
(5) 14

40. The number of grams in one kilogram is

(1) .001
(2) .01
(3) .1
(4) 10
(5) 1000

41. An appliance store gives a 15% discount off the list price of all of its merchandise. An additional 30% reduction of the store price is made for the purchase of a floor model. A television set which has a list price of \$300 and is a floor model sells for

(1) \$210.00
(2) \$228.50
(3) \$178.50
(4) \$165.00
(5) \$135.00

42. Mrs. Jones wishes to buy 72 ounces of canned beans for the least possible cost. Which of the following should she buy?

(1) Six 12-ounce cans at 39¢ per can
(2) Seven 10-ounce cans at 34¢ per can
(3) Three 24-ounce cans at 79¢ per can
(4) Two 25-ounce cans at 62¢ per can
(5) Five 13-ounce cans at 37¢ per can

43. The distance from point A to point B is

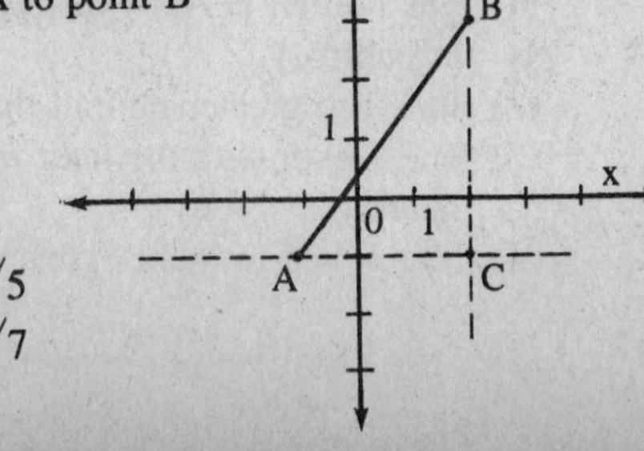

(1) 3
(2) 5
(3) 6
(4) $\sqrt{5}$
(5) $\sqrt{7}$

44. If $x^2 - x - 6 = 0$, then x is equal to

(1) 3 only
(2) -2 only
(3) -3 or 2
(4) 3 or -2
(5) -3 or -2

45. Which quantity is *not* equal to $75(32 + 88)$?

(1) $75 \cdot 32 + 75 \cdot 88$
(2) $(75 \cdot 32) + 88$
(3) $75(88 + 32)$
(4) $(88 + 32) \cdot 75$
(5) $88 \cdot 75 + 32 \cdot 75$

46. In a certain boys' camp, 30% of the boys are from New York State and 20% of these are from New York City. What percent of the boys in the camp are from New York City?

(1) 60
(2) 50
(3) 20
(4) 10
(5) 6

47. If 1 ounce is approximately equal to 28 grams, then 1 pound is approximately equal to

(1) 250 grams
(2) 350 grams
(3) 450 grams
(4) 550 grams
(5) 650 grams

48. Which fraction is equal to .25%?

(1) $\frac{1}{400}$
(2) $\frac{1}{40}$
(3) $\frac{1}{4}$
(4) $\frac{5}{2}$
(5) $\frac{50}{2}$

Questions 49 and 50 refer to the table below:

VALUE OF PROPERTY STOLEN — 1977 and 1978
LARCENY

CATEGORY	1977 Number of Offenses	1977 Value of Stolen Property	1978 Number of Offenses	1978 Value of Stolen Property
Pocket-picking	20	$ 1,950	10	$ 950
Purse-snatching	175	5,750	120	12,050
Shoplifting	155	7,950	225	17,350
Automobile thefts	1040	127,050	860	108,000
Thefts of automobile accessories	1135	34,950	970	24,400
Bicycle thefts	355	8,250	240	6,350
All other thefts	1375	187,150	1300	153,150

49. Of the total number of larcenies reported in 1977, automobile thefts accounted for, most nearly,

(1) 5%
(2) 15%
(3) 25%
(4) 50%
(5) 65%

50. Of the following, the category which had the largest reduction in value of stolen property from 1977 to 1978 was

(1) pocket-picking
(2) automobile thefts
(3) shoplifting
(4) bicycle thefts
(5) purse-snatching

Answer Key for Examination II Mathematics for GED Candidates

1. (1)	11. (2)	21. (5)	31. (5)	41. (3)
2. (3)	12. (1)	22. (3)	32. (5)	42. (1)
3. (3)	13. (5)	23. (5)	33. (2)	43. (2)
4. (1)	14. (3)	24. (3)	34. (2)	44. (4)
5. (4)	15. (3)	25. (3)	35. (4)	45. (2)
6. (3)	16. (5)	26. (2)	36. (1)	46. (5)
7. (2)	17. (4)	27. (3)	37. (3)	47. (3)
8. (2)	18. (4)	28. (3)	38. (1)	48. (1)
9. (2)	19. (3)	29. (5)	39. (3)	49. (3)
10. (2)	20. (3)	30. (2)	40. (5)	50. (2)

Solutions for Examination II

1. $$\begin{array}{r} 304{,}802 \\ -\ 212{,}810 \\ \hline 91{,}992 \end{array}$$

Answer: **(1)** 91,992

2. From 9:30 AM to 3 PM is $5\frac{1}{2}$ hours.

$$\$4 \cdot 5\tfrac{1}{2} = \$22$$

Answer: **(3)** $22.00

3. $$\begin{array}{rl} .010 & \text{(3 decimal places)} \\ \times\ .001 & \text{(3 decimal places)} \\ \hline .000010 & \text{(6 decimal places)} \end{array}$$

The final zero may be dropped:

$$.000010 = .00001$$

Answer: **(3)** .00001

4. When $x = 3$, $2x + y = 7$ becomes

$$2 \cdot 3 + y = 7$$

Solve for y:

$$\begin{array}{rcl} 6 + y &=& 7 \\ -6 & & -6 \\ \hline y &=& 1 \end{array}$$

Answer: **(1)** 1

5. Let s = Paul's annual salary

$$5\% \text{ of } s = \$750$$
$$.05s = \$750$$
$$\frac{.05s}{.05} = \frac{\$750}{.05}$$
$$s = \$15{,}000$$

Answer: **(4)** $15,000

6. $(-6) + (-2)(-3) = (-6) + (+6)$ First multiply, then add.
$= 0$

Answer: **(3)** 0

7. To round 825.6347 to the nearest hundredth, consider 4, the digit in the thousandths place. Since it is less than 5, drop all digits to the right of the hundredths place.

   825.6347 = 825.63 to the nearest hundredth

   Answer: **(2)** 825.63

8. Point P has coordinates $x = -3$ and $y = 2$.

   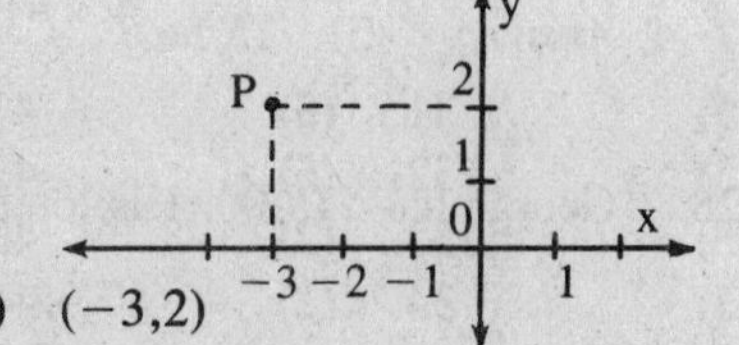

   Answer: **(2)** (−3,2)

9. The boy buys oranges for 10¢ each (30¢ ÷ 3). He sells them for 12¢ each (60¢ ÷ 5). Therefore his profit is 2¢ per orange. He must sell 50¢ ÷ 2¢ = 25 oranges for a 50¢ profit.

   Answer: **(2)** 25

10. $V = \frac{1}{3}\pi r^2 h$, r = 3″, h = 14″

    $V = \frac{1}{\cancel{3}} \cdot \frac{22}{\cancel{7}} \cdot \cancel{3} \cdot 3 \cdot \cancel{14}$ (cancelling: 3 → 1, 7 → 1, 3 → 1, 14 → 2)

    $V = \frac{22 \cdot 3 \cdot 2}{1}$

    $V = 132$

    Answer: **(2)** 132 cubic inches

11. Find the difference between each choice and 5:

5.000	5.005	5.01	5.1	5.105
−4.985	−5.000	−5.00	−5.0	−5.000
.015	.005	.01	.1	.105
		= .010	= .100	

    The smallest difference is .005, therefore 5.005 is closer than the other choices to 5.

    Answer: **(2)** 5.005

12. 4 yards = 4 · 3 feet = 12 feet
    12 feet ÷ 3 = 4 feet per piece

    Answer: **(1)** 4 feet

13. Area of a triangle = $\frac{1}{2}$ · base · height

    The height, which is 4, is drawn to base AB, which is 6.

    $$\text{Area} = \tfrac{1}{2} \cdot 6 \cdot 4 = 3 \cdot 4 = 12$$

    Answer: **(5)** 12

14. If 1 dozen rolls costs \$1.50, each roll costs

    \$1.50 ÷ 12 = \$.125

    Then 30 rolls will cost

    30(\$.125) = \$3.75

    Answer: **(3)** \$3.75

15. $P = 2(a + b)$

    If $a = 3$ and $b = 4$,

    $$P = 2(3 + 4) = 2(7) = 14$$

    Answer: **(3)** 14

16. The sum of the angles of a triangle is 180°. The two given angles total 80°.

    180° − 80° = 100°

    The third angle is 100°.

    Answer: **(5)** 100

17. Let x represent the actual distance between towns, then write a proportion:

    $$\frac{x}{50} = \frac{4\frac{1}{2}}{1\frac{1}{2}}$$

    $$\frac{x}{50} = 3 \qquad \left(\text{Since } 4\tfrac{1}{2} \div 1\tfrac{1}{2} = 3\right)$$

    $$x = 150$$

    Answer: **(4)** 150 miles

18. Each of the numbers in the sequence is a perfect square:

    1, 4, 9, 16, 25, . . .
    $1^2, 2^2, 3^2, 4^2, 5^2, \ldots$

    The next number is 6^2, or 36.

    Answer: **(4)** 36

19. | | |
|---|---|
| \$3.32 | total charge |
| − 1.24 | charge for first five ounces |
| \$2.08 | charge for additional weight at \$.16 per ounce |

    2.08 ÷ .16 = 13

    5 ounces + 13 ounces = 18 ounces
    = 1 pound 2 ounces
    = $1\frac{1}{8}$ pound

    Answer: **(3)** $1\frac{1}{8}$ pound

20. $2\frac{1}{2} \cdot 3 = \frac{5}{2} \cdot 3$
$= \frac{15}{2}$
$= 7\frac{1}{2}$

Answer: **(3)** $7\frac{1}{2}$

21. Substitute the coordinates of each point in each equation. Only $y = -x$ is satisfied by the coordinates of the points:

$(-2,2)$: $2 = -(-2)$
$(3,-3)$: $-3 = -(3)$

Answer: **(5)** $y = -x$

22. 12 feet = 4 yards
15 feet = 5 yards
4 yards · 5 yards = 20 square yards

$20.80 per square yard
× 20 square yards
$416.00

Answer: **(3)** $416.00

23. There are three pounds of peanuts.

$\frac{3}{5} = .60 = 60\%$

Answer: **(5)** 60

24. January: .4
February: 1.4
March: 2.2
Total: 4.0

Answer: **(3)** 4.0

25. April: 2.8
May: 2.4
June: 1.4
Total: 6.6

Average: $6.6 \div 3 = 2.2$

Answer: **(3)** 2.2

26. The rainfall in June was 1.4 inches, the same as the rainfall in February.

Answer: **(2)**

27. Let x represent the person's height, and write a proportion:

$\frac{5}{x} = \frac{8}{24}$

$\frac{5}{x} = \frac{1}{3}$

$x = 15$ (Cross multiply)

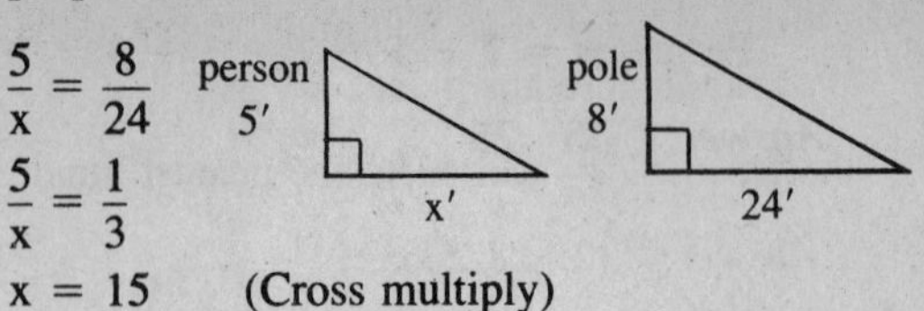

Answer: **(3)** 15 feet

28. George has $15.00. His total purchase is:

$7.32
+ 1.68
$9.00

He will have $15.00 − $9.00 = $6.00 left.

Answer: **(3)** $6.00

29. There are 52 cards in a deck, of which 13 are clubs. The probability of picking a club is $\frac{13}{52} = \frac{1}{4}$.

Answer: **(5)** $\frac{1}{4}$

30. It takes Jack 48 minutes ÷ 6 miles = 8 minutes for each mile. At that rate it will take him $15 \cdot 8 = 120$ minutes for 15 miles.

120 minutes = 2 hours

Answer: **(2)** 2 hours

31.
$$\begin{aligned} 2x - 7 &= 3 \\ +7 &\quad +7 \\ 2x &= 10 \\ x &= 5 \end{aligned}$$

If $x = 5$, $3x + 1 = 3 \cdot 5 + 1 = 16$

Answer: **(5)** 16

32. The increase in sales was

$17,394,683
− 13,382,675
$ 4,012,008

The average yearly increase over 4 years was

$4,012,008 ÷ 4 = $1,003,002

Answer: **(5)** $1,003,002

33. The perimeter is the sum of all the sides of the figure. ABED is a rectangle, so side ED = 7.

Perimeter = 7 + 5 + 7 + 3 + 3 = 25

Answer: **(2)** 25

34. Kilometer is used to measure long distances.

Answer: **(2)** kilometer

35. I = p · r · t, where I = interest, p = principal, r = rate, t = time in years.

If p = $200, r = 12%, and t = 2 years,

I = ($200)(12%)(2)
= ($200)(.12)(2)
= $48.00

Answer: **(4)** $48

36. There are 6 symbols representing men and 8 symbols representing women. The ratio is 6:8, or 3:4.

Answer: **(1)** 3:4

37.

Monday:	300 children	300 adults
Thursday:	500 children	200 adults
Total:	800 children	500 adults

Total revenue = 800($.50) + 500($1.50)
= $400 + $750
= $1150

Answer: **(3)** $1150

38. There are 30 symbols in all. Each symbol represents 100 people.

30 · 100 = 3000

Answer: **(1)** 3000

39. Try each choice:

(1) 5 5¢ stamps = 25¢
100¢ − 25¢ = 75¢
= exactly 25 3¢ stamps

(2) 8 5¢ stamps = 40¢
100¢ − 40¢ = 60¢
= exactly 20 3¢ stamps

(3) 9 5¢ stamps = 45¢
100¢ − 45¢ = 55¢
= 18 3¢ stamps and 1¢ change

(4) 11 5¢ stamps = 55¢
100¢ − 55¢ = 45¢
= exactly 15 3¢ stamps

(5) 14 5¢ stamps = 70¢
100¢ − 70¢ = 30¢
= exactly 10 3¢ stamps

In choice (3), exactly $1.00 can *not* be spent.

Answer: **(3)** 9

40. 1 kilogram = 1000 grams

Answer: **(5)** 1000

41. The price after the 15% discount is

85% of $300 = .85($300)
= $255

The price after the 30% discount is

70% of $255 = .70($255)
= $178.50

Answer: **(3)** $178.50

42. Only choices (1) and (3) represent 72 ounces.

Choice (1): 6($.39) = $2.34
Choice (3): 3($.79) = $2.37

Answer: **(1)** Six 12-ounce cans at 39¢ per can.

43. ABC is a right triangle.
AC = 3
BC = 4

Using the Pythagorean Theorem,

$$(AB)^2 = 3^2 + 4^2 = 9 + 16 = 25$$
$$AB = \sqrt{25} = 5$$

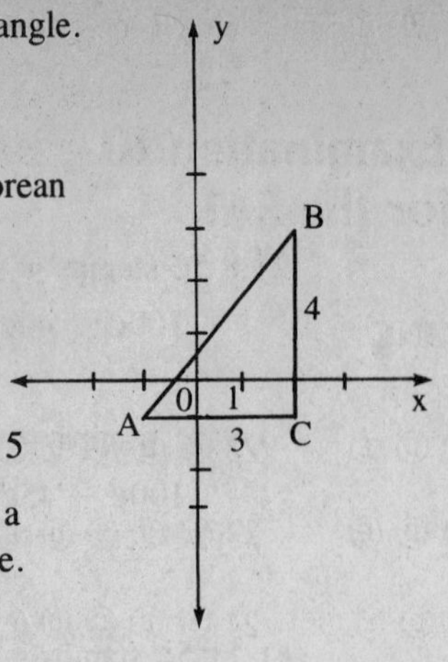

Note that ABC is a 3-4-5 right triangle.

Answer: **(2)** 5

44.
$$x^2 - x - 6 = 0$$
$$(x - 3)(x + 2) = 0$$
$$x - 3 = 0 \quad x + 2 = 0$$
$$x = 3 \quad x = -2$$

An alternate method is to substitute each given answer into the equation to determine which are solutions. For example, in choice (3), $x = -3$

$$(-3)^2 - (-3) - 6 = 9 + 3 - 6 = 6 \neq 0$$

Therefore $x = -3$ is not a solution of $x^2 - x - 6 = 0$.

Answer: **(4)** 3 or −2

45. Choices (1), (3), (4) and (5) are all examples of the commutative and distributive properties. The quantity in choice (2) is not equal to $75(32 + 88)$.

Answer: **(2)** $(75 \cdot 32) + 88$

46.
$$20\% \text{ of } 30\% = (.20)(.30) = .06 = 6\%$$

Answer: **(5)** 6%

47.
1 pound = 16 ounces
$(16)(28) = 448$

Answer: **(3)** 450 grams

48.
$$.25\% = .0025 = \frac{25}{10000} = \frac{1}{400}$$

Answer: **(1)** $\frac{1}{400}$

49. Total larcenies in 1977:

$$\begin{array}{r} 20 \\ 175 \\ 155 \\ 1040 \\ 1135 \\ 355 \\ +\ 1375 \\ \hline 4255 \end{array}$$

$$\frac{\text{Auto thefts}}{\text{Total}} = \frac{1040}{4255} = .24 \text{ (approximately)} = 24\%$$

Answer: **(3)** 25%

50. Pocket-picking: 1950 − 950 = 1000 reduction

Auto thefts: 127,050 − 108,000 = 19,050 reduction

Shoplifting: increased

Bicycle thefts: 8250 − 6350 = 1900 reduction

Purse-snatching: increased

Answer: **(2)** automobile thefts

Answer Sheet for Examination III
Mathematics for the SAT

PART ONE

1 Ⓐ Ⓑ Ⓒ Ⓓ Ⓔ	8 Ⓐ Ⓑ Ⓒ Ⓓ Ⓔ	15 Ⓐ Ⓑ Ⓒ Ⓓ Ⓔ	22 Ⓐ Ⓑ Ⓒ Ⓓ Ⓔ	29 Ⓐ Ⓑ Ⓒ Ⓓ Ⓔ
2 Ⓐ Ⓑ Ⓒ Ⓓ Ⓔ	9 Ⓐ Ⓑ Ⓒ Ⓓ Ⓔ	16 Ⓐ Ⓑ Ⓒ Ⓓ Ⓔ	23 Ⓐ Ⓑ Ⓒ Ⓓ Ⓔ	30 Ⓐ Ⓑ Ⓒ Ⓓ Ⓔ
3 Ⓐ Ⓑ Ⓒ Ⓓ Ⓔ	10 Ⓐ Ⓑ Ⓒ Ⓓ Ⓔ	17 Ⓐ Ⓑ Ⓒ Ⓓ Ⓔ	24 Ⓐ Ⓑ Ⓒ Ⓓ Ⓔ	31 Ⓐ Ⓑ Ⓒ Ⓓ Ⓔ
4 Ⓐ Ⓑ Ⓒ Ⓓ Ⓔ	11 Ⓐ Ⓑ Ⓒ Ⓓ Ⓔ	18 Ⓐ Ⓑ Ⓒ Ⓓ Ⓔ	25 Ⓐ Ⓑ Ⓒ Ⓓ Ⓔ	32 Ⓐ Ⓑ Ⓒ Ⓓ Ⓔ
5 Ⓐ Ⓑ Ⓒ Ⓓ Ⓔ	12 Ⓐ Ⓑ Ⓒ Ⓓ Ⓔ	19 Ⓐ Ⓑ Ⓒ Ⓓ Ⓔ	26 Ⓐ Ⓑ Ⓒ Ⓓ Ⓔ	33 Ⓐ Ⓑ Ⓒ Ⓓ Ⓔ
6 Ⓐ Ⓑ Ⓒ Ⓓ Ⓔ	13 Ⓐ Ⓑ Ⓒ Ⓓ Ⓔ	20 Ⓐ Ⓑ Ⓒ Ⓓ Ⓔ	27 Ⓐ Ⓑ Ⓒ Ⓓ Ⓔ	34 Ⓐ Ⓑ Ⓒ Ⓓ Ⓔ
7 Ⓐ Ⓑ Ⓒ Ⓓ Ⓔ	14 Ⓐ Ⓑ Ⓒ Ⓓ Ⓔ	21 Ⓐ Ⓑ Ⓒ Ⓓ Ⓔ	28 Ⓐ Ⓑ Ⓒ Ⓓ Ⓔ	35 Ⓐ Ⓑ Ⓒ Ⓓ Ⓔ

PART TWO

1 Ⓐ Ⓑ Ⓒ Ⓓ Ⓔ	6 Ⓐ Ⓑ Ⓒ Ⓓ Ⓔ	11 Ⓐ Ⓑ Ⓒ Ⓓ Ⓔ	16 Ⓐ Ⓑ Ⓒ Ⓓ Ⓔ	21 Ⓐ Ⓑ Ⓒ Ⓓ Ⓔ
2 Ⓐ Ⓑ Ⓒ Ⓓ Ⓔ	7 Ⓐ Ⓑ Ⓒ Ⓓ Ⓔ	12 Ⓐ Ⓑ Ⓒ Ⓓ Ⓔ	17 Ⓐ Ⓑ Ⓒ Ⓓ Ⓔ	22 Ⓐ Ⓑ Ⓒ Ⓓ Ⓔ
3 Ⓐ Ⓑ Ⓒ Ⓓ Ⓔ	8 Ⓐ Ⓑ Ⓒ Ⓓ Ⓔ	13 Ⓐ Ⓑ Ⓒ Ⓓ Ⓔ	18 Ⓐ Ⓑ Ⓒ Ⓓ Ⓔ	23 Ⓐ Ⓑ Ⓒ Ⓓ Ⓔ
4 Ⓐ Ⓑ Ⓒ Ⓓ Ⓔ	9 Ⓐ Ⓑ Ⓒ Ⓓ Ⓔ	14 Ⓐ Ⓑ Ⓒ Ⓓ Ⓔ	19 Ⓐ Ⓑ Ⓒ Ⓓ Ⓔ	24 Ⓐ Ⓑ Ⓒ Ⓓ Ⓔ
5 Ⓐ Ⓑ Ⓒ Ⓓ Ⓔ	10 Ⓐ Ⓑ Ⓒ Ⓓ Ⓔ	15 Ⓐ Ⓑ Ⓒ Ⓓ Ⓔ	20 Ⓐ Ⓑ Ⓒ Ⓓ Ⓔ	25 Ⓐ Ⓑ Ⓒ Ⓓ Ⓔ

EXAMINATION III
MATHEMATICS FOR THE SAT

60 questions — 1 hour

This examination provides an excellent review of all that you have learned in MATHEMATICS SIMPLIFIED AND SELF-TAUGHT. In addition, it is an accurate indicator of your chances for success in the Mathematical Ability Sections of the Scholastic Aptitude Test. The questions are similar to the ones that appear on the SAT in number and level of difficulty. The timing and type of question are just what you may expect on the actual test. When you have completed the entire examination, check your work with the detailed solutions provided at the end of the test.

PART ONE: 35 questions — 30 minutes

DIRECTIONS: The problems in this section are to be solved using any available space on the page itself for scratch work. When the problem has been worked out, indicate the one appropriate answer on the answer sheet.

The information which follows should be helpful in determining the correct answers for some of the problems.

Circle of radius r: Area = πr^2; Circumference = $2\pi r$.
The number of degrees in a circle is 360.
The number of degrees in a straight angle is 180.

Definition of symbols:

$\|\|$	is parallel to	$>$	is greater than
$\leqslant$	is less than or equal to	$<$	is less than
$\geqslant$	is greater than or equal to	$\perp$	is perpendicular to
$\angle$ or $\measuredangle$	angle	$\triangle$	triangle

Triangle: The sum of the measures in degrees of the angles in a triangle is 180. The angle BDC is a right angle; therefore,

(1) the area of triangle ABC $= \dfrac{AC \times BD}{2}$

(2) $AB^2 = AD^2 + DB^2$

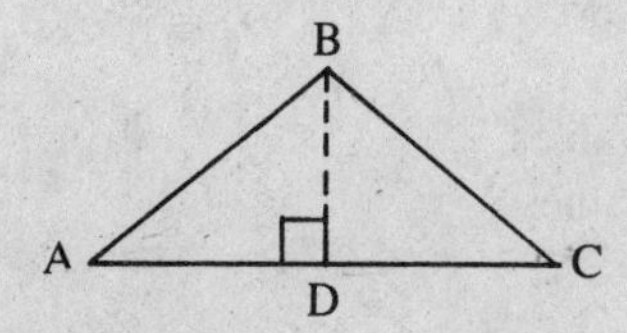

Note: The figures which accompany these problems are drawn as accurately as possible *unless* stated otherwise in specific problems. Again, unless stated otherwise, all figures lie in the same plane. All numbers used in these problems are real numbers.

1. The value of $\frac{2}{3}(\frac{3}{4} - \frac{1}{3})$ is

(A) $\frac{1}{6}$
(B) $\frac{5}{18}$
(C) $\frac{1}{2}$
(D) 1
(E) $\frac{4}{3}$

2. If the area of a square of side x is 5, what is the area of a square of side 3x?

(A) $3\sqrt{5}$
(B) $9\sqrt{5}$
(C) 15
(D) 45
(E) 75

3. If $1 + \frac{1}{t} = \frac{t+1}{t}$, what does t equal?

(A) only +1
(B) only +1 or −1
(C) only 0 or +1
(D) any number except 0
(E) no number

4. In the figure at right, if AB || DE and AC = BC, then x =

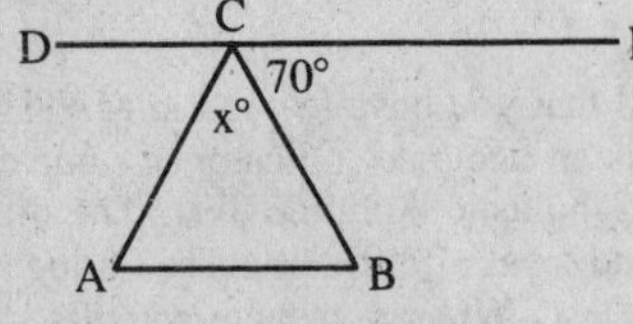

(A) 110°
(B) 90°
(C) 70°
(D) 50°
(E) 40°

5. A proper fraction is unchanged in value if both numerator and denominator are

(A) increased by the same number
(B) decreased by the same number
(C) divided by the same number
(D) raised to the second power
(E) replaced by their square roots

6. A man travels a distance of 20 miles at 60 miles per hour and returns over the same route at 40 miles per hour. What is his average rate for the round trip in miles per hour?

(A) 50
(B) 48
(C) $47\frac{1}{2}$
(D) $33\frac{1}{3}$
(E) 30

7. For what values of n is −n equal to −(−n)?

(A) no value
(B) 0
(C) all negative values
(D) all positive values
(E) all values

8. The figure at right is composed of nine equal squares and has a total area of 144. The perimeter of the figure is

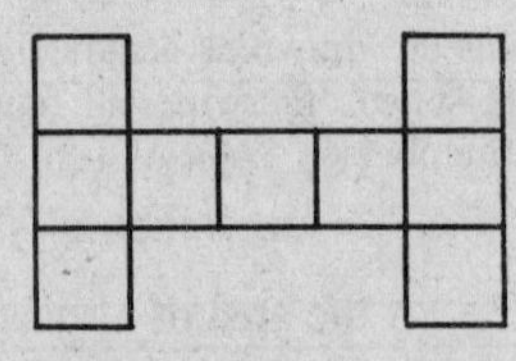

(A) 40
(B) 64
(C) 72
(D) 80
(E) 144

9. If $m^2 + 1 = 7$, then $m^4 + 2m^2 =$

(A) 36
(B) 38
(C) 48
(D) 49
(E) 58

10. John can shovel the driveway in x minutes. After he has worked for y minutes, what part of the driveway is still unshoveled?

(A) $\frac{x}{y}$
(B) $\frac{y}{x}$
(C) $\frac{x-y}{y}$
(D) $\frac{y-x}{x}$
(E) $\frac{x-y}{x}$

11. What is the diameter of the largest circle which can be drawn tangent to both AB and CD in parallelogram ABCD at right?

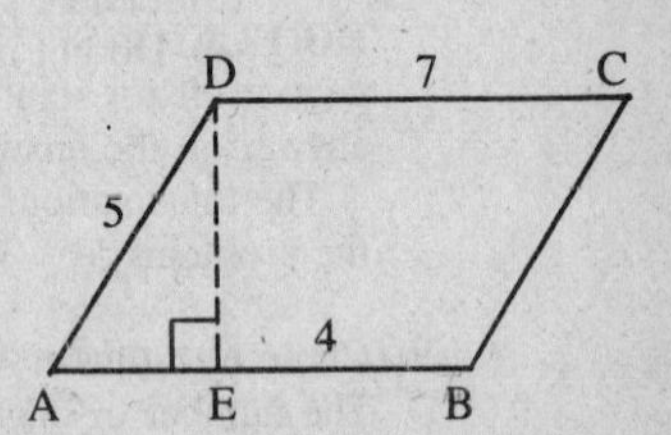

(A) 1
(B) 2
(C) 3
(D) 4
(E) 5

AD = 5
CD = 7
EB = 4

Questions 12 and 13 refer to the following information:

Consider five geometric figures: an isosceles triangle with base equal to its altitude = a, a square with side = a, a circle with radius = a, a regular hexagon with each side = a, and a semicircle with diameter = 2a.

12. Which figure has the greatest area?

(A) triangle
(B) square
(C) circle
(D) hexagon
(E) semicircle

13. Which figure has the greatest perimeter?

(A) triangle
(B) square
(C) circle
(D) hexagon
(E) semicircle

14. If $d = m - \frac{50}{m}$, and m is a positive number, then as m increases in value, d must

(A) increase in value
(B) decrease in value
(C) remain unchanged
(D) increase until m = 50, then decrease
(E) decrease until m = 50, then increase

15. Ruth and Sarah together have \$20. Sarah and Teresa together have \$40. Teresa and Ruth together have \$30. How much money does Ruth alone have?

(A) \$5
(B) \$10
(C) \$15
(D) \$20
(E) \$25

DIRECTIONS: For questions 16–35, compare two quantities, one in Column A and one in Column B, and mark your answer sheet:

(A) if the quantity is greater in Column A
(B) if the quantity is greater in Column B
(C) if both quantities are equal
(D) if no comparison can be made with the given information

Notes: (1) Information concerning one or both of the compared quantities will be centered above the two columns in some of the questions.
(2) Symbols that appear in both columns represent the same thing in Column A as in Column B.
(3) Letters such as x, n, and k are symbols for real numbers.
(4) Do not mark choice (E), as there are only four choices.

	COLUMN A	COLUMN B
16.	The product of any two negative numbers	The product of any positive number and any negative number

17. $3x = 100$

	COLUMN A	COLUMN B
17.	$\frac{2}{3}x$	50

18.

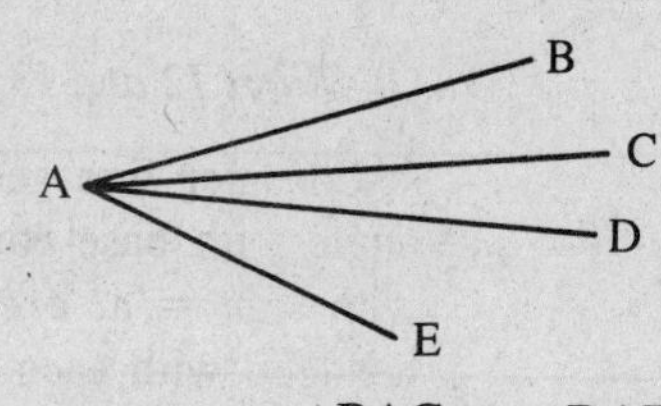

$\angle BAC > \angle DAE$

	COLUMN A	COLUMN B
18.	$\angle BAD$	$\angle CAE$
19.	$a(a + b)$	$a^2 + ab$
20.	The greatest prime number x such that $x^2 < 84$.	The greatest odd integer x such that $x^2 < 84$.

21.

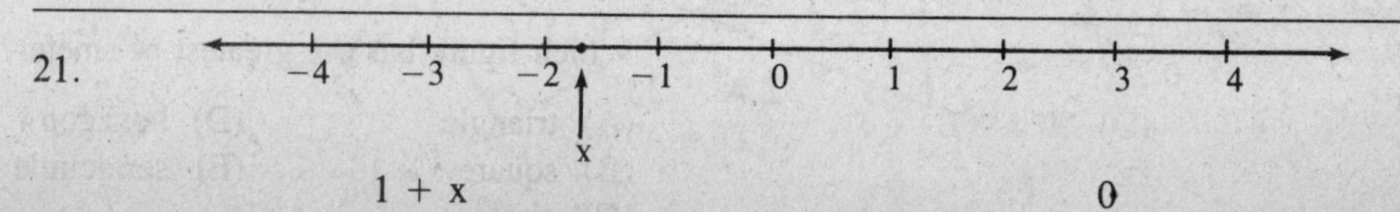

	COLUMN A	COLUMN B
21.	$1 + x$	0

	Column A		Column B
22.		$x^2 = 81$ $y^2 = 100$	
	x		y
23.	The difference between $\frac{9}{8}$ and $\frac{3}{5}$		.5

24.

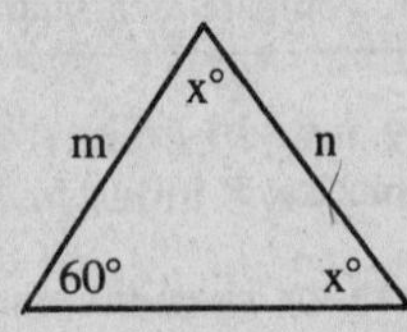

Column A	Column B
m	n

25.

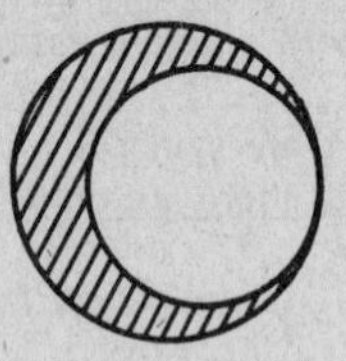

radius of large circle = 7
radius of small circle = 5

Column A	Column B
area of small circle	area of shaded portion

	Column A		Column B
26.		$a - b = -1$ $-b - a = -3$	
	a		b
27.	$\sqrt{.09}$		$(.09)^2$
28.	The average of 101, 103, 105, 107		The average of 102, 104, 106
29.		a, b, and c are positive integers and $a > b > c$	
	$\frac{a}{b}$		$\frac{c}{b}$

30.

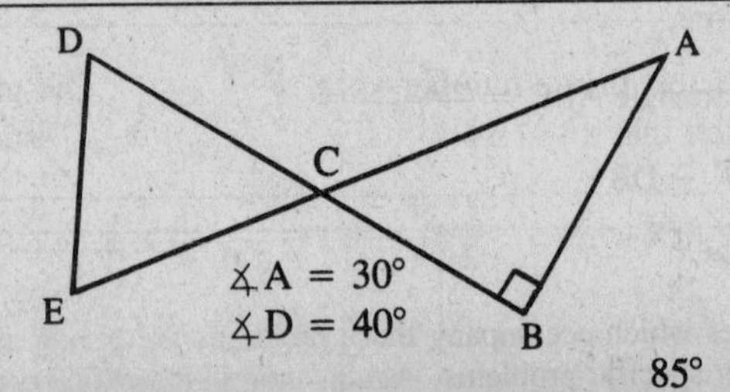

$\measuredangle A = 30°$
$\measuredangle D = 40°$

Column A	Column B
$\measuredangle E$	85°

	COLUMN A		COLUMN B
31.		$x + 1 > 0$ $y - 1 > 0$	
	x		y
32.		$A \square B = \left(\frac{1}{A}\right)^2 - \frac{1}{B}$	
	$\frac{1}{2} \square \frac{1}{4}$		$\frac{1}{4} \square \frac{1}{2}$
33.		City A is 10 miles from City B City B is 8 miles from City C	
	The distance from City A to City C		18 miles
34.		$\frac{a}{a+b} = \frac{c}{c+d}$	
	bc		ad
35.		$A < 0$ and $B < 0$	
	A^2		B^3

PART TWO: 25 questions — 30 minutes

DIRECTIONS: The problems in this section are to be solved using any available space on the page itself for scratch work. When the problem has been worked out, indicate the one appropriate answer on the answer sheet.

The information which follows should be helpful in determining the correct answers for some of the problems.

Circle of radius r: Area = πr^2; Circumference = $2\pi r$.
The number of degrees in a circle is 360.
The number of degrees in a straight angle is 180.

Definition of symbols:

$\parallel$	is parallel to	$>$	is greater than
$\leqslant$	is less than or equal to	$<$	is less than
$\geqslant$	is greater than or equal to	$\perp$	is perpendicular to
$\angle$ or $\measuredangle$	angle	$\triangle$	triangle

Triangle: The sum of the measures in degrees of the angles in a triangle is 180. The angle BDC is a right angle; therefore,

(1) the area of triangle ABC $= \frac{AC \times BD}{2}$

(2) $AB^2 = AD^2 + DB^2$

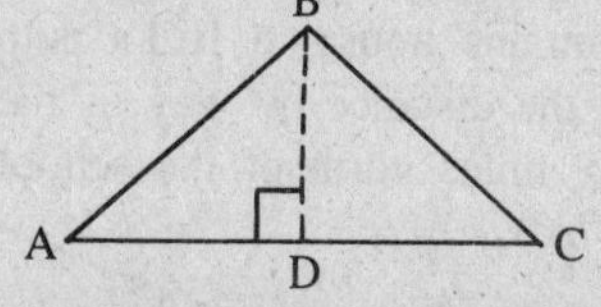

Note: The figures which accompany these problems are drawn as accurately as possible *unless* stated otherwise in specific problems. Again, unless stated otherwise, all figures lie in the same plane. All numbers used in these problems are real numbers.

1. The quantity $c - d$ must be negative if

(A) $d > 0$
(B) $c < d$
(C) $c < 0$
(D) $d < c$
(E) $0 > d$

2. If $x + y = 5$ and $xy = 6$, then $x^2 + y^2 =$

(A) 11
(B) 12
(C) 13
(D) 25
(E) 30

Questions 3 and 4 refer to the following definition:

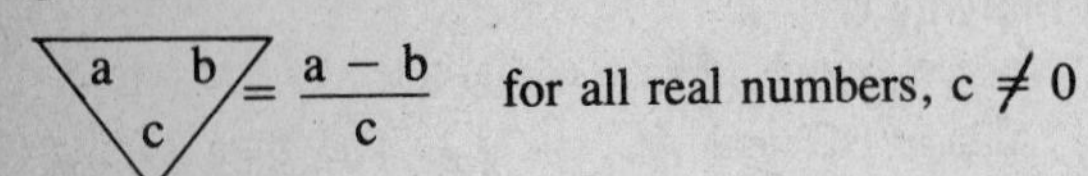

for all real numbers, $c \neq 0$

3. 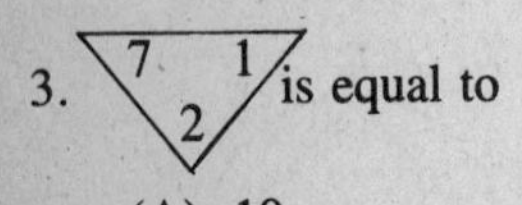

is equal to

(A) 10
(B) 7
(C) 4
(D) 3
(E) 0

4. 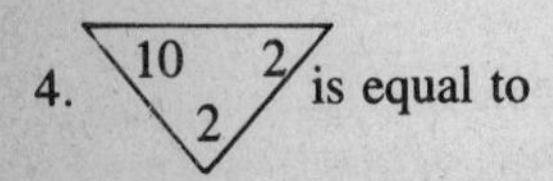

is equal to

(A) 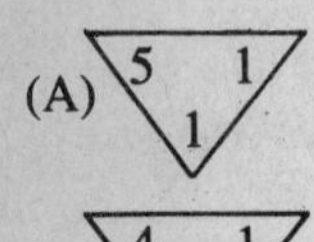

(B) and (C) 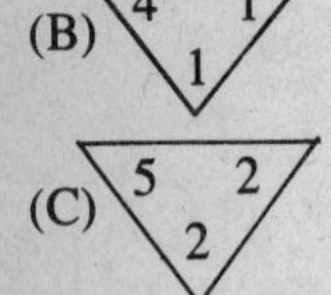

(D) 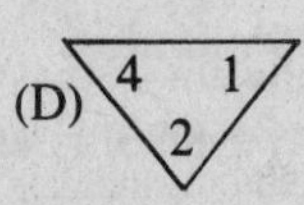

(E) 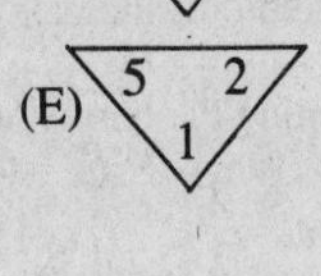

5. How many of the numbers between 200 and 400 begin or end with 3?

(A) 20
(B) 40
(C) 100
(D) 110
(E) 180

6. Peter lives 12 miles west of school and Bill lives north of the school. Peter finds that the direct distance from his house to Bill's is 6 miles shorter than the distance by way of the school. How many miles north of the school does Bill live?

(A) 6
(B) 9
(C) 10
(D) $6\sqrt{2}$
(E) $10\sqrt{3}$

7. The distance s in feet that a body falls in t seconds is given by the formula $s = 16t^2$. If a body has been falling for 4 seconds, how far will it fall in the 5th second?

(A) 64 feet
(B) 80 feet
(C) 144 feet
(D) 176 feet
(E) 256 feet

8. According to the graph at right, the life expectancy in the United States in 1925 was about what percent of the average life expectancy in 1825?

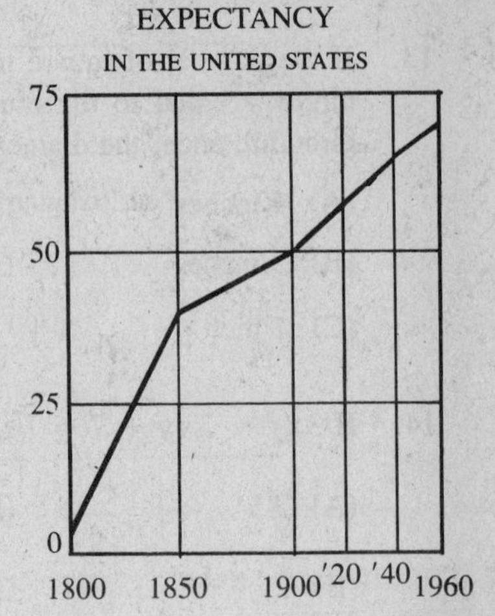

(A) 150
(B) 175
(C) 200
(D) 250
(E) 300

9. Which of the following pairs of equations represent perpendicular lines?

(A) $x = y$
$x = y + 1$
(B) $y = -1$
$y = 1$
(C) $x = 1$
$x = -1$
(D) $x + y = 3$
$x + y = -3$
(E) $x = -1$
$y = 1$

10. If the angles of a triangle are in the ratio 2:3:7, the triangle is

(A) acute
(B) isosceles
(C) obtuse
(D) right
(E) equilateral

Questions 11 and 12 refer to the following unit block:

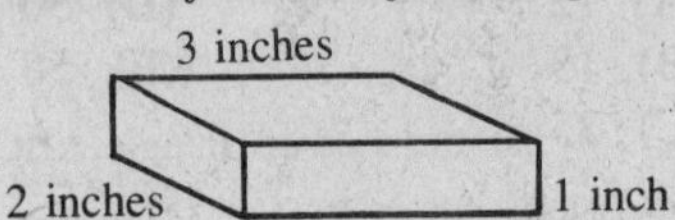

11. How many whole blocks are needed to construct a solid cube of minimum size?

(A) 6
(B) 18
(C) 36
(D) 48
(E) 215

12. How many blocks, arranged as shown in the diagram below, and including half blocks, would be required to build a wall 1 yard long, 2 inches deep and $\frac{1}{2}$ foot high?

    (A) 36
    (B) 72
    (C) 144
    (D) 216
    (E) 432

13. If the number of square inches in the area of a circle is equal to the number of inches in its circumference, the diameter of the circle is

    (A) 4 inches
    (B) 2 inches
    (C) 1 inch
    (D) π inches
    (E) $\frac{\pi}{2}$ inches

14. If $3xy - 2xy + 7 = 10$, then $x =$

    (A) 3
    (B) $3y$
    (C) $3xy$
    (D) $\frac{y}{3}$
    (E) $\frac{3}{y}$

15. In the figure, M is the midpoint of the base of parallelogram RSTU. What is the ratio of the area of triangle RSM to the area of the parallelogram?

    (A) 1:2
    (B) 1:4
    (C) 1:3
    (D) 2:5
    (E) 2:7

16. A gear 50 inches in diameter turns a smaller gear 25 inches in diameter. If the larger gear makes 15 revolutions, how many revolutions does the smaller gear make in that time?

    (A) 12
    (B) 15
    (C) 25
    (D) 30
    (E) 50

17. If $3x = 2k$ and $5y = 8k$, then the ratio $x:y$ is equal to

    (A) 5:12
    (B) 5:8
    (C) 12:5
    (D) 8:5
    (E) 3:4

18. If m and n are both integers greater than 2, then of the following, the fraction which is largest in value is

    (A) $\frac{m}{2n}$
    (B) $\frac{m}{2n-4}$
    (C) $\frac{m}{2n-2}$
    (D) $\frac{m}{2n+2}$
    (E) $\frac{m}{2n+4}$

19. In the figure, AB is a diameter of the circle, $AC = BC$ and $AC = \sqrt{2}$. The area of the circle is

    (A) $\frac{\pi\sqrt{2}}{2}$
    (B) π
    (C) $\pi\sqrt{2}$
    (D) 2π
    (E) 4π

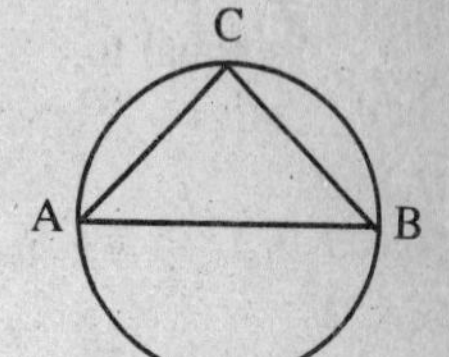

20. Below are shown three views of a single cube. What symbol is on the bottom of figure 3?

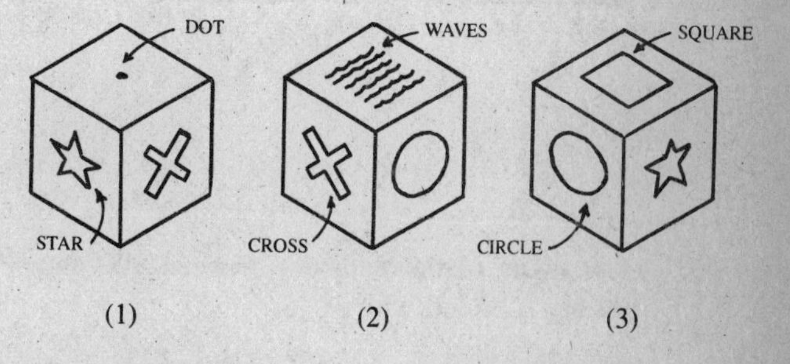

    (A) cross
    (B) dot
    (C) star
    (D) waves
    (E) circle

21. A store reduced the price of eggs from $1.00 per dozen to 2 dozen for $1.60. What was the percent decrease per dozen?

    (A) 16
    (B) 20
    (C) $25\frac{1}{2}$
    (D) $37\frac{1}{2}$
    (E) 40

22. A student has an average of 78% in four of his major subjects. What is the least grade he must get in his fifth major subject if he is to average 80% in all five majors?

(A) 82
(B) 84
(C) 86
(D) 88
(E) 90

23. The length of each side of a square is $\frac{2x}{3} + 1$. The perimeter of the square is

(A) $\frac{8x + 4}{3}$
(B) $\frac{8x + 12}{3}$
(C) $\frac{2x}{3} + 4$
(D) $\frac{2x}{3} + 16$
(E) $\frac{4x}{3} + 2$

24. If one angle of a triangle is three times a second angle, and the third angle is 20° more than the second angle, the second angle is

(A) 32°
(B) 34°
(C) 40°
(D) 50°
(E) 60°

25. If one side of a square is increased by 3 and an adjacent side is decreased by 3, a rectangle is formed whose area is 40. What is the area of the original square?

(A) 31
(B) 34
(C) 37
(D) 43
(E) 49

Answer Key to Examination III
Mathematics for the SAT

PART ONE

1. **(B)**	8. **(D)**	15. **(A)**	22. **(D)**	29. **(A)**
2. **(D)**	9. **(C)**	16. **(A)**	23. **(A)**	30. **(B)**
3. **(D)**	10. **(E)**	17. **(B)**	24. **(C)**	31. **(D)**
4. **(E)**	11. **(D)**	18. **(A)**	25. **(A)**	32. **(B)**
5. **(C)**	12. **(C)**	19. **(C)**	26. **(B)**	33. **(D)**
6. **(B)**	13. **(C)**	20. **(B)**	27. **(A)**	34. **(C)**
7. **(B)**	14. **(A)**	21. **(B)**	28. **(C)**	35. **(A)**

PART TWO

1. **(B)**	6. **(B)**	11. **(C)**	16. **(D)**	21. **(B)**
2. **(C)**	7. **(C)**	12. **(B)**	17. **(A)**	22. **(D)**
3. **(D)**	8. **(E)**	13. **(A)**	18. **(B)**	23. **(B)**
4. **(A)**	9. **(E)**	14. **(E)**	19. **(B)**	24. **(A)**
5. **(D)**	10. **(C)**	15. **(B)**	20. **(A)**	25. **(E)**

Solutions to Examination III

PART ONE

1.
$$\tfrac{2}{3}\left(\tfrac{3}{4} - \tfrac{1}{3}\right) = \tfrac{2}{3}\left(\tfrac{9}{12} - \tfrac{4}{12}\right)$$
$$= \frac{\overset{1}{\cancel{2}}}{3} \cdot \frac{5}{\underset{6}{\cancel{12}}}$$
$$= \tfrac{5}{18}$$

Answer: **(B)** $\frac{5}{18}$

2. The area of a square of side x is x^2.

$$x^2 = 5$$

The area of a square of side 3x is $(3x)^2$.

$$(3x)^2 = 9x^2$$
$$= 9(5)$$
$$= 45$$

Answer: **(D)** 45

3. Multiply both sides of the equation by t:

$$t\left(1 + \frac{1}{t}\right) = t\left(\frac{t+1}{t}\right)$$
$$t + 1 = t + 1$$

This is an identity which is true for all values of t. However, 0 must be excluded because of the division by t in the original equation.

Answer: **(D)** Any number except 0

4. $AC = BC$, therefore $\measuredangle A = \measuredangle B$
$DE \parallel AB$, therefore $\measuredangle B = \measuredangle ECB$ and
$\measuredangle A = \measuredangle DCA$

Then $\measuredangle B = 70°$, $\measuredangle A = 70°$, and $\measuredangle DCA = 70°$.

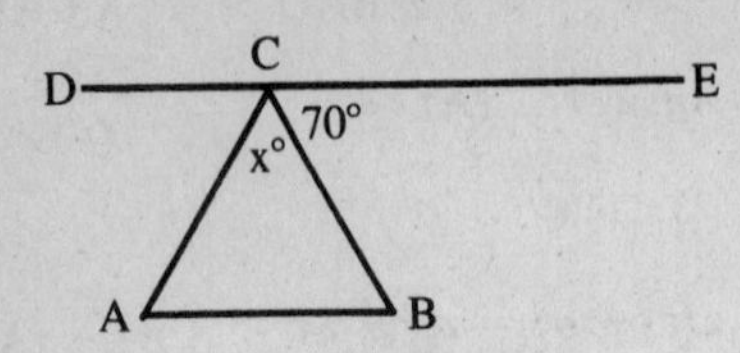

$$\measuredangle DCA + x° + 70° = 180°$$
$$70° + x° + 70° = 180°$$
$$x° + 140° = 180°$$
$$x° = 40°$$

Answer: **(E)** 40°

5. If both numerator and denominator of a fraction are divided by the same number, the value of the fraction remains unchanged.

Answer: **(C)** divided by the same number

6. $\text{Time} = \dfrac{\text{Distance}}{\text{Rate}}$

Time for 20 miles at 60 miles per hour $= \frac{20}{60}$
$= \frac{1}{3}$ hour

Time for 20 miles at 40 miles per hour $= \frac{20}{40}$
$= \frac{1}{2}$ hour

Total distance = 40 miles

Total time $= \frac{1}{3} + \frac{1}{2} = \frac{5}{6}$ hour

$$\text{Rate} = \frac{\text{Total distance}}{\text{Total time}}$$
$$= \frac{40 \text{ miles}}{\frac{5}{6} \text{ hour}}$$
$$= 40 \div \tfrac{5}{6} \text{ miles per hour}$$
$$= 40 \cdot \tfrac{6}{5} \text{ miles per hour}$$
$$= 48 \text{ miles per hour}$$

Answer: **(B)** 48

7. If $-n = -(-n)$,
then $-n = +n$, which is true only if $n = 0$.

Answer: **(B)** 0

8. Area of each square $= 144 \div 9 = 16$
Side of each square $= 4$

The perimeter consists of 20 sides:

Perimeter $= 20 \cdot 4 = 80$

Answer: **(D)** 80

9. $m^2 + 1 = 7$, therefore $m^2 = 6$

$$m^4 + 2m^2 = (m^2)^2 + 2m^2$$
$$= (6)^2 + 2(6)$$
$$= 36 + 12$$
$$= 48$$

Answer: **(C)** 48

10. John shovels $\frac{1}{x}$ of the driveway each minute. After y minutes, he has $x - y$ minutes left to work, and the fraction $\frac{x-y}{x}$ is still unshoveled.

Answer: **(E)** $\dfrac{x-y}{x}$

11. The largest circle has diameter equal to DE. In parallelogram ABCD, AB = CD; therefore, if CD = 7 and EB = 4, AE = 3. Triangle AED is a 3–4–5 right triangle, with DE = 4.

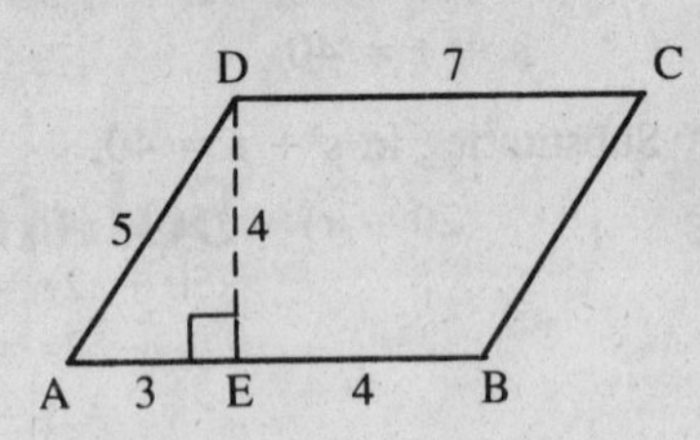

Answer: **(D)** 4

12. and 13.

Compare the circle, semicircle and regular hexagon to the square and the triangle:

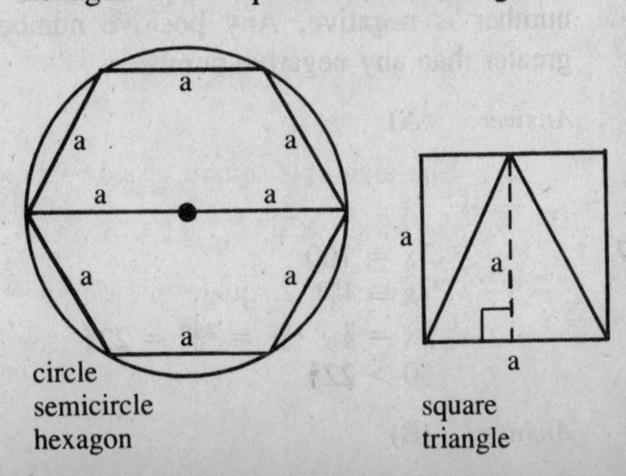

The circle has a larger area and perimeter than either the semicircle or the hexagon.

The square has a larger area and perimeter than the triangle.

	Area	*Perimeter*
Circle	πa^2	$2\pi a$
Square	a^2	$4a$

The circle has the largest area and perimeter.

12. *Answer:* **(C)** circle

13. *Answer:* **(C)** circle

14. As m increases, $\frac{50}{m}$ decreases and $m - \frac{50}{m}$ increases.

Answer: **(A)** increase in value

15. Let r = Ruth's money
s = Sarah's money
t = Teresa's money

Then $r + s = 20$, or $s = 20 - r$
$t + r = 30$, or $t = 30 - r$
$s + t = 40$

Substituting in $s + t = 40$,

$$(20 - r) + (30 - 4) = 40$$
$$50 - 2r = 40$$
$$-2r = -10$$
$$r = 5$$

Answer: **(A)** \$5

16. The product of two negative numbers is positive. The product of a positive number and a negative number is negative. Any positive number is greater than any negative number.

Answer: **(A)**

17.
$$3x = 100$$
$$x = \tfrac{100}{3}$$
$$\tfrac{2}{3}x = \tfrac{2}{3} \cdot \tfrac{100}{3} = \tfrac{200}{9} = 22\tfrac{2}{9}$$
$$50 > 22\tfrac{2}{9}$$

Answer: **(B)**

18. Adding the same quantity to both sides of an inequality maintains the order of inequality.

$$\begin{array}{rcr} \angle BAC & > & \angle DAE \\ +\ \angle CAD & & +\ \angle CAD \\ \hline \angle BAD & > & \angle CAE \end{array}$$

Answer: **(A)**

19. $a(a + b) = a^2 + ab$

Answer: **(C)**

20. The greatest prime integer whose square is less than 84 is 7. The greatest odd integer whose square is less than 84 is 9.

Answer: **(B)**

21. If x is between -2 and -1, $1 + x$ must be between -1 and 0 and is therefore less than 0.

Answer: **(B)**

22. Solutions for $x^2 = 81$ are 9 and -9.
Solutions for $y^2 = 100$ are 10 and -10.
No comparison can be made between x and y.

Answer: **(D)**

23.
$$\tfrac{9}{8} = 1.125$$
$$\tfrac{3}{5} = .600$$
$$.525 \text{ difference}$$
$$.525 > .5$$

Answer: **(A)**

24.
$$60° + x° + x° = 180°$$
$$60° + 2x° = 180°$$
$$2x° = 120°$$
$$x° = 60°$$

The triangle is equilateral, therefore $m = n$.

Answer: **(C)**

25. Area of large circle $= 49\pi$
Area of small circle $= 25\pi$
Area of shaded portion $= 49\pi - 25\pi = 24\pi$

Area of small circle > area of shaded portion.

Answer: **(A)**

26. Rearrange the equations, then add:

$$\begin{aligned} a - b &= -1 \\ -a - b &= -3 \\ \hline -2b &= -4 \\ b &= 2 \\ a - 2 &= -1 \\ a &= 1 \end{aligned}$$

$$b > a$$

Answer: **(B)**

27.
$$\sqrt{.09} = .3$$
$$(.09)^2 = .0081$$
$$.3 > .0081$$

Answer: **(A)**

28.
$$\frac{101 + 103 + 105 + 107}{4} = \frac{416}{4} = 104$$

$$\frac{102 + 104 + 106}{3} = \frac{312}{3} = 104$$

Answer: **(C)**

29. Dividing both sides of an inequality by a positive number maintains the order of inequality:

$$a > c$$
$$\frac{a}{b} > \frac{c}{b}$$

Answer: **(A)**

30.

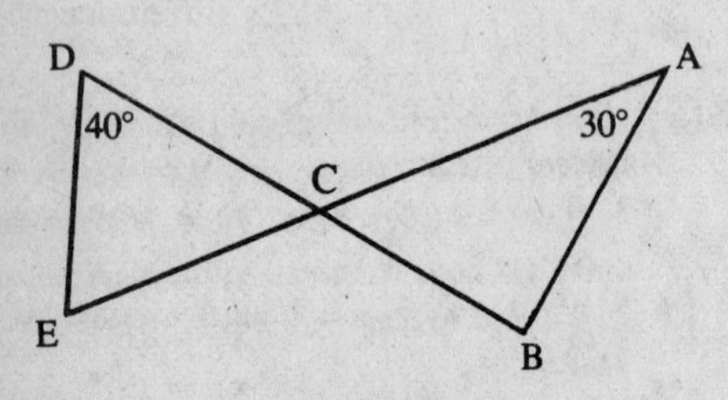

$$\measuredangle ACB + 30° + 90° = 180°$$
$$\measuredangle ACB + 120° = 180°$$
$$\measuredangle ACB = 60°$$

$\measuredangle ACB$ and $\measuredangle DCE$ are vertical angles, therefore $\measuredangle DCE = 60°$.

$$\measuredangle E + 40° + 60° = 180°$$
$$\measuredangle E + 100° = 180°$$
$$\measuredangle E = 80°$$

Answer: **(B)**

31.
$$x + 1 > 0 \qquad y - 1 > 0$$
$$x > -1 \qquad y > 1$$

Some specific values of x which satisfy the first inequality are greater than, some are equal to, and some are less than, specific values of y which satisfy the second inequality.

Answer: **(D)**

32.
$$\tfrac{1}{2} \square \tfrac{1}{4} = \left(\frac{1}{\frac{1}{2}}\right)^2 - \frac{1}{\frac{1}{4}} = 2^2 - 4 = 0 \qquad \tfrac{1}{4} \square \tfrac{1}{2} = \left(\frac{1}{\frac{1}{4}}\right)^2 - \frac{1}{\frac{1}{2}} = 4^2 - 2 = 14$$

Answer: **(B)**

33. If the three cities lie on the same straight line, the distance from A to C is 18 miles. If they do not, the distance is less than 18 miles.

Answer: **(D)**

34.
$$\begin{aligned} \frac{a}{a+b} &= \frac{c}{c+d} \\ a(c + d) &= c(a + b) \\ ac + ad &= ac + bc \\ -ac & \quad -ac \\ \hline ad &= bc \end{aligned}$$

The product of the means is equal to the product of the extremes.

Answer: **(C)**

35. The square of a negative number is positive. The cube of a negative number is negative. Any positive number is greater than any negative number.

Answer: **(A)**

PART TWO

1.
$$\begin{aligned} c - d &< 0 \\ + d & \quad +d \\ \hline c &< d \end{aligned}$$

Answer: **(B)** $c < d$

2. If $x + y = 5$, $(x + y)^2 = 25$

$$\begin{aligned} x^2 + 2xy + y^2 &= 25 \\ x^2 + 2(6) + y^2 &= 25 \\ x^2 + 12 + y^2 &= 25 \\ x^2 + y^2 &= 13 \end{aligned}$$

Answer: **(C)** 13

3. 7 1 2 $= \frac{7-1}{2} = 3$

Answer: **(D)** 3

4. 10 2 2 $= \frac{10-2}{2} = \frac{5-1}{1} =$ 5 1 1

Answer: **(A)** 5 1 1

5. All the numbers from 300 to 399 begin with 3. There are 100 of these.

Also all the numbers such as 203, 213, . . . , 293 end with 3. There are 10 of these. Therefore there are 110 numbers between 200 and 400 that begin or end with 3.

Answer: **(D)** 110

6. Let x = the direct distance from Peter to Bill
y = the distance from school to Bill

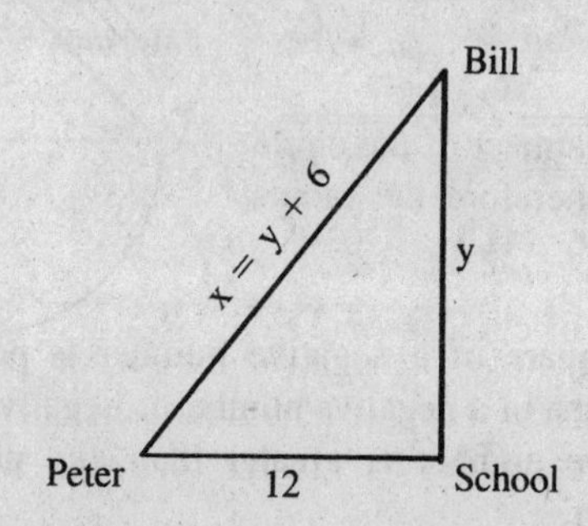

Then x is 6 miles less than y + 12

$$x = y + 12 - 6$$
$$x = y + 6$$

Then

$$(y+6)^2 = y^2 + 12^2$$
$$y^2 + 12y + 36 = y^2 + 144$$
$$12y + 36 = 144$$
$$12y = 108$$
$$y = 9$$

Answer: **(B)** 9

7. In 4 seconds, $s = 16 \cdot 4^2 = 256$
In 5 seconds, $s = 16 \cdot 5^2 = 400$

$$400 - 256 = 144$$

Answer: **(C)** 144

8. According to the graph, in 1925, the life expectancy was about 60 years, while in 1825 it was about 20 years. 60 is 300% of 20.

Answer: **(E)** 300

9. $x = -1$ represents a line parallel to the y-axis.
$y = \ \ 1$ represents a line parallel to the x-axis.

These lines are perpendicular to each other.

Answer: **(E)** $x = -1$
$y = 1$

10. Represent the angles as 2x, 3x and 7x.

Then

$$2x + 3x + 7x = 180°$$
$$12x = 180°$$
$$x = 15°$$

The angles of the triangle are

$$2x = 30°$$
$$3x = 45°$$
$$7x = 105°$$

The triangle is obtuse.

Answer: **(C)** obtuse

11. One edge of the minimum cube must be 6 inches, the lowest common multiple of 1, 2 and 3. The volume of the cube will be 6^3, or 216 cubic inches, which is equal to 36 blocks, since 216 cubic inches ÷ 6 cubic inches per block = 36 blocks.

Answer: **(C)** 36

12. The volume of the wall will be $36 \cdot 2 \cdot 6 = 432$ cubic inches.

432 cubic inches ÷ 6 cubic inches per block = 72 blocks

Answer: **(B)** 72

13. The area of the circle is πr^2 and its circumference is $2\pi r$. If the area equals the circumference,

$$\pi r^2 = 2\pi r$$

Divide both sides by πr:

$$\frac{\pi r^2}{\pi r} = \frac{2\pi r}{\pi r}$$
$$r = 2$$

If the radius is 2 inches, the diameter is 4 inches.

Answer: **(A)** 4 inches

14.
$$3xy - 2xy + 7 = 10$$
$$xy + 7 = 10$$
$$xy = 3$$
$$x = \frac{3}{y}$$

Answer: **(E)** $\frac{3}{y}$

15. The height h of triangle RSM is also the height of parallelogram RSTU.

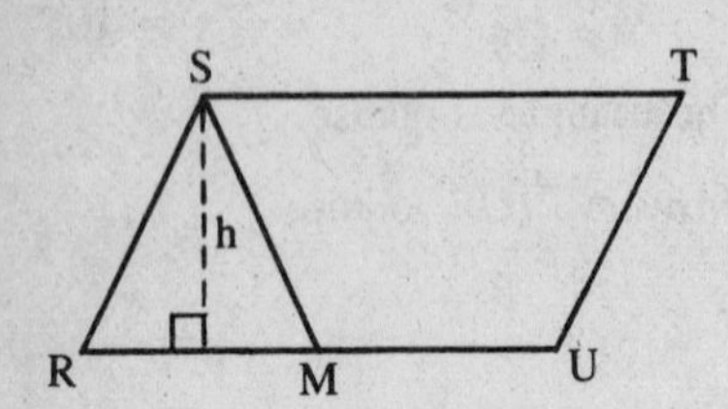

$$\text{Area of triangle RSM} = \tfrac{1}{2} \cdot h \cdot (RM)$$
$$\text{Area of parallelogram RSTU} = h \cdot (RU)$$
$$= h \cdot (2RM)$$
$$= 2h(RM)$$

$$\frac{\text{Area of triangle RSM}}{\text{Area of parallelogram RSTU}} = \frac{\frac{1}{2}h(RM)}{2h(RM)}$$
$$= \frac{\frac{1}{2}}{2} = \frac{1}{4}$$

Answer: **(B)** 1:4

16. The smaller gear makes two revolutions for each revolution of the larger gear. If the larger gear makes 15 revolutions, the smaller gear makes 30.

Answer: **(D)** 30

17. If $3x = 2k$, $x = \frac{2k}{3}$. If $5y = 8k$, $y = \frac{8k}{5}$.

$$\frac{x}{y} = \frac{2k}{3} \div \frac{8k}{5} = \frac{\overset{1}{\cancel{2k}}}{3} \cdot \frac{5}{\underset{4}{\cancel{8k}}} = \frac{5}{12}$$

Answer: **(A)** 5:12

18. The restriction that m and n are greater than 2 insures that the fractions are positive. If two or more positive fractions have the same numerator, the fraction which is largest in value has the smallest denominator. The smallest denominator is $2n - 4$; therefore the largest fraction is $\frac{m}{2n - 4}$.

Answer: **(B)** $\frac{m}{2n - 4}$

19. ∡C is an inscribed angle which intercepts a semicircle (180°) and is therefore a right angle.

$$(AB)^2 = (\sqrt{2})^2 + (\sqrt{2})^2$$
$$= 2 + 2$$
$$= 4$$
$$AB = \sqrt{4} = 2$$

The diameter of the circle is 2; therefore the radius is 1.

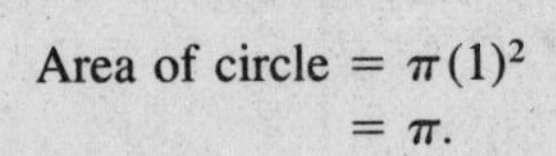

$$\text{Area of circle} = \pi(1)^2$$
$$= \pi.$$

Answer: **(B)** π

20. The cube may be unfolded as shown:

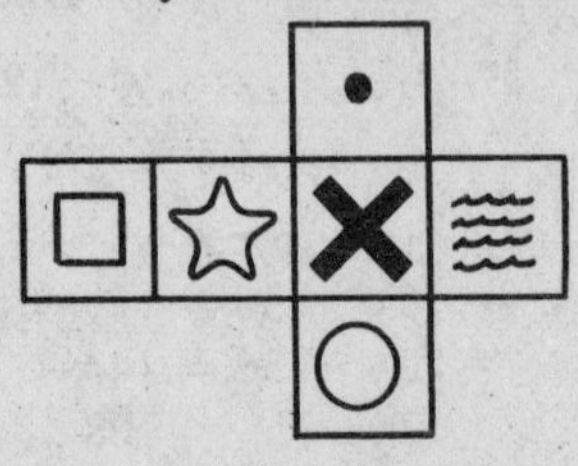

The symbol opposite the square is the cross.

Answer: **(A)** cross

21. The old price was \$1.00 per dozen. The new price is \$.80 per dozen. The decrease is \$.20 per dozen.

$$\frac{.20}{1.00} = 20\%$$

Answer: **(B)** 20

22. $\frac{\text{Sum of 4 majors}}{4} = 78$

$\text{Sum of 4 majors} = 78 \cdot 4 = 312$

$\frac{\text{Sum of 5 majors}}{5} = 80$

$\frac{312 + \text{fifth major}}{5} = 80$

$312 + \text{fifth major} = 80 \cdot 5 = 400$

$\text{fifth major} = 400 - 312 = 88$

Answer: **(D)** 88

23. $4\left(\frac{2x}{3} + 1\right) = \frac{8x}{3} + 4$

$= \frac{8x}{3} + \frac{12}{3}$

$= \frac{8x + 12}{3}$

Answer: **(B)** $\frac{8x + 12}{3}$

24. Let s = second angle
then $3s$ = first angle
$s + 20$ = third angle

$s + 3s + (s + 20°) = 180°$

$5s + 20° = 180°$

$5s = 160°$

$s = 32°$

Answer: **(A)** 32°

25.

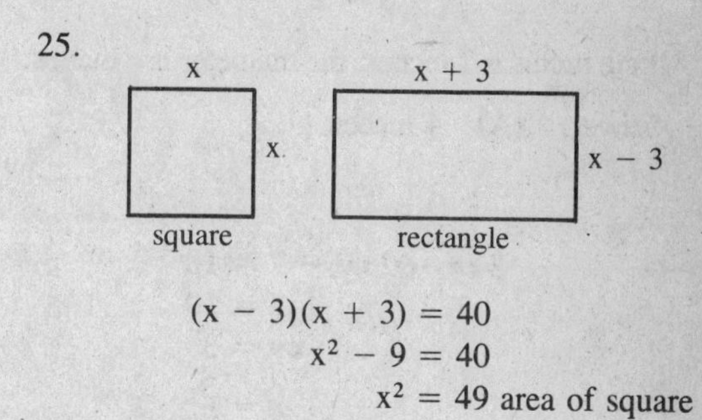

$(x - 3)(x + 3) = 40$

$x^2 - 9 = 40$

$x^2 = 49$ area of square

Answer: **(E)** 49